Probabilistic Intelligence in Geotechnical Engineering

By
Jean Paul

Table of Contents

CHAPTER 1
Introduction

1.1 Background

In geotechnical engineering, the assessment and modelling of subsurface conditions are inherently uncertain due to the natural variability of soil properties, limitations in site investigation methods, and constraints related to time, cost, and accessibility. This uncertainty can have significant implications for design safety, reliability, and cost-effectiveness. Despite these challenges, traditional deterministic approaches continue to dominate in practice. These methods typically use fixed input parameters, rely heavily on empirical correlations, and apply safety factors to account for variability. While straightforward to implement, they often fail to represent the full extent of uncertainty in geotechnical behaviour, leading either to over-conservative designs or to the underestimation of risk.

As a response to this limitation, probabilistic methods have gained attention in recent decades, offering a more rigorous way to capture uncertainty and improve the reliability of geotechnical analyses. Among these, Bayesian methods have emerged as particularly powerful. The Bayesian framework allows engineers to systematically update their understanding of soil behaviour as more data becomes available, integrating prior knowledge with new evidence in a coherent, mathematically grounded way. This makes Bayesian approaches especially well-suited for geotechnical applications, where data from boreholes, in-situ tests, laboratory experiments, and past project experiences must be synthesised to inform key engineering decisions. The core strength of Bayesian methods lies in their ability to explicitly model both aleatory uncertainty (arising from natural variability) and epistemic uncertainty (arising from lack of knowledge or limited data). Unlike deterministic models, which produce single-point estimates, Bayesian approaches yield full probability distributions for parameters and predictions. This allows for richer interpretation, better risk management, and more robust design choices. Bayesian models can also be updated iteratively as new site data becomes available, making them ideal for projects where phased investigations or staged construction are involved. Advancements in computational methods have further expanded the applicability of Bayesian inference in geotechnical contexts. Tools like Markov Chain Monte Carlo (MCMC), hierarchical Bayesian modelling, and Bayesian neural networks now make it feasible to apply these concepts even to complex systems. These tools allow for more sophisticated probabilistic analyses, such as regional correlation modelling, performance prediction under variable loading, and reliability assessment in seismic or groundwater-related problems. Bayesian machine learning, a natural extension of Bayesian statistics, has added further value to this landscape. Unlike conventional machine learning models that focus on point estimates, Bayesian machine learning represents model parameters and predictions as probability distributions. This means the models not only provide an answer but also tell you how confident they are in that answer. Such uncertainty-aware predictions are highly valuable in geotechnical engineering, where data is often sparse, costly, and site-specific.

Another important advantage is the ability of Bayesian models to incorporate prior knowledge. In geotechnical engineering, prior knowledge might come from site history, regional soil profiles, past failures, or even expert judgment. Bayesian modelling formalises this by allowing the user to define prior distributions that reflect existing beliefs, which are then updated as new data becomes available. This is particularly useful in scenarios where limited field testing has been conducted, but some historical or regional data exists.

Bayesian methods also offer principled approaches to model comparison and selection. By evaluating the posterior probabilities of different models, engineers can choose the one most consistent with observed data and prior expectations. This reduces reliance on ad hoc model selection techniques, such as cross-validation, which may not always be statistically sound, especially in small-sample settings common in geotechnical work. Bayesian optimisation further complements this toolkit by efficiently exploring design or model parameter spaces. When the objective function is expensive to evaluate or unknown, as is often the case in foundation design or site selection, Bayesian optimisation acts as a surrogate to guide decision-making. It is especially powerful for tuning machine learning models and has demonstrated success in improving the performance of ensemble learning algorithms used for predictive tasks in geotechnical applications.

In sum, Bayesian approaches provide a consistent, flexible, and data-efficient alternative to deterministic or black-box models. They bridge the gap between statistical rigour and engineering practicality, offering tools that are not only powerful but also interpretable, a critical requirement for geotechnical engineers who must justify their decisions in safety-critical contexts.

1.2 Motivation

Despite the strong theoretical advantages of Bayesian methods, their practical use in geotechnical engineering remains limited. Several key factors explain this gap between academic promise and field adoption. One is the technical complexity associated with these methods. Bayesian analysis requires a solid grounding in probability theory and statistical inference, and in many cases, a working knowledge of programming and computational methods. Most geotechnical engineers are not formally trained in these areas, making it difficult to implement Bayesian approaches without significant additional learning.

Another barrier is the lack of widespread support in commercial geotechnical software. Many standard packages used in practice do not currently include Bayesian capabilities, which limits accessibility for practising engineers. Developing user-friendly tools that integrate Bayesian frameworks into familiar software platforms would be a crucial step toward broader adoption. A further practical challenge is the specification of prior distributions, which can seem subjective or arbitrary to engineers unfamiliar with probabilistic reasoning. However, advances in hierarchical modelling and empirical Bayes approaches are making it easier to derive priors from data, helping to standardise and justify their use.

In choosing the focus areas of this research, several motivations converged. First, the selected topics parameter estimation (shear wave velocity -SPT correlation), bearing capacity prediction, groutability classification, and liquefaction susceptibility are core to geotechnical design. They represent real-world problems that engineers routinely deal with, and they involve significant uncertainty that cannot be adequately addressed through deterministic methods alone. Second, in each of these areas, a large number of empirical and deterministic models already exist, often with contradictory results. These models are typically built on site-specific data, which means they may not transfer well across regions or conditions. This leads to confusion and inconsistency in practice. A more generalisable, data-driven approach that still respects geotechnical principles was clearly needed.

Third, while the literature in machine learning has begun to explore some of these areas, most models are either opaque or poorly generalised. They work well on one dataset but fail when applied to another. Many are not interpretable, which limits their usefulness in engineering decision-making. In contrast, Bayesian models offer transparency, flexibility, and adaptability, especially when data is sparse or noisy. Importantly, geotechnical engineering is one of the few fields where we are fortunate to have a significant body of high-quality field test data, such as plate load tests, CPT data, and SPT results. While gathering new data is expensive, this existing information provides a valuable resource. The goal of this thesis is to leverage that data more effectively to build models that reduce the need for excessive field testing without compromising reliability. By doing so, we not only reduce cost but also make better use of the information already available from past projects. What unifies the research across all chapters is a consistent theme: improving geotechnical prediction and decision-making under uncertainty. Each chapter explores a different application, but all are tied together by a commitment to probabilistic thinking and Bayesian reasoning. The models developed in this work are intended to be practical, interpretable, and transferable, not just statistically impressive, but genuinely useful to engineers in the field.

Ultimately, this thesis aims to contribute to a shift in geotechnical engineering practice from deterministic simplifications to probabilistic understanding; from isolated, site-specific models to frameworks that learn from broader datasets; and from opaque "black box" predictions to transparent, data-informed decisions. By demonstrating how Bayesian methods can be applied to real geotechnical problems in meaningful ways, this research hopes to help pave the way toward more reliable, efficient, and data-smart engineering practice.

1.3 Organisation of the Thesis

This thesis is structured into seven chapters, each systematically addressing the research objectives and contributing to a coherent investigation of advanced probabilistic and machine learning techniques in geotechnical engineering.

Chapter 1: Introduction

This chapter provides a comprehensive overview of the motivation, significance, and scope of the research. It introduces the central themes of uncertainty quantification, Bayesian modelling, and data-driven methods in geotechnical engineering, and presents the four primary objectives of the thesis. The need for probabilistic tools and machine learning in improving predictive accuracy and interpretability in soil mechanics and foundation engineering is discussed.

Chapter 2: Literature Review

This chapter surveys the existing body of work across the four key focus areas of the study: uncertainty quantification and various Bayesian methods in geotechnical engineering. It explores research on the correlation between shear wave velocity and SPT blow counts, the prediction of bearing pressure using machine learning, the modelling of soil groutability, and the estimation of liquefaction hazard through Bayesian logistic regression. The chapter critically examines empirical, statistical, and AI-based approaches used in these domains, discusses their strengths and limitations, and identifies the research gap that this thesis aims to address.

Chapter 3: Bayesian Multilevel Modelling for the Vs and N relationship in Sandy Soil

This chapter develops a Bayesian multilevel model to characterise the shear wave velocity and SPT blow counts relationship across multiple geographic regions. The statistical framework accounts for both global trends and regional variations using hierarchical structures. The results confirm the validity of a constant slope with variable intercepts and demonstrate the power of Bayesian inference in geotechnical parameter estimation.

Chapter 4: Bayesian-Optimised Ensemble Models for Spread Footings on Clay

In this chapter, ensemble learning techniques—AdaBoost and GBoost—are optimised using Bayesian Optimisation Algorithms (BOA) to predict bearing pressure based on plate load test data. The chapter details the data curation, model development, hyperparameter tuning, and performance benchmarking against traditional methods. The GBoost-BOA model is shown to provide the highest accuracy and generalisation.

Chapter 5: Hybrid Bayesian Network Approach for Soil Groutability

This chapter introduces a novel hybrid Bayesian network model that incorporates both continuous and discrete variables to predict soil groutability. The model structure is learned using MCMC and hill-climbing algorithms, and performance is evaluated against discrete

BN, SVM, and Naïve Bayes classifiers. Emphasis is placed on high recall and practical applicability in groutability assessment.

Chapter 6: Bayesian Logistic Regression for Soil Liquefaction Using CPT Data

In this chapter, a Bayesian logistic regression model is developed for classifying liquefied and non-liquefied cases using field CPT data. Statistically guided transformations improve linearity and prediction accuracy.

Chapter 7: Summary and Conclusions

The final chapter synthesises the key findings from each research component, highlighting their contributions to geotechnical modelling and decision-making. It discusses the broader implications and potential engineering applications of each model, identifies methodological limitations, and suggests directions for future research.

CHAPTER

2

Literature Review

2.1 Uncertainty

Geotechnical engineering is inherently uncertain due to the complex nature of soil and rock materials, which are shaped by diverse geological processes that are not entirely observable or replicable. This uncertainty is not merely a byproduct of limited data but is deeply embedded in the material behaviour and methods of analysis used in the field. Barbour and Krahn (2004) highlighted that geotechnical engineers often underemphasize prediction accuracy, noting that it is not uncommon for computed and measured responses to differ by an order of magnitude.

The field has long recognised the dominance of uncertainty. Curran and Hammah (2006), in discussing software development for geomechanical applications, famously stated, “uncertainty is king.” They emphasised that due to the heterogeneous and discontinuous nature of geologic materials—formed under unknown and variable conditions—single-value predictions are rarely accurate, necessitating the explicit accommodation of uncertainty in analysis tools.

Phoon (2017) extended this perspective by arguing that geotechnical engineering remains as much an art as a science, partly because conventional models often overlook the “known unknowns” and “unknown unknowns” intrinsic to subsurface conditions. Soils, being porous and formed by weathering, exhibit significant spatial heterogeneity even in uniform layers. Tang (1984) identified three principal sources of uncertainty: inherent spatial variability, systematic errors arising from testing processes, and estimation errors due to limited sampling. Phoon and Kulhawy (1999) further refined this framework by distinguishing among (1) inherent variability of soil properties, (2) measurement error, and (3) transformation uncertainty associated with empirical correlations. They advocated for treating each source separately and combining them using a probabilistic framework, such as the second-moment method, to evaluate the overall uncertainty in geotechnical parameters. The coefficient of variation (COV) is commonly employed to quantify this variability.

Christian et al. (1994) applied probabilistic techniques to assess slope stability, categorising uncertainty into random data scatter and systematic error. They demonstrated that variability in design inputs, such as shear strength and pore pressure, can significantly affect reliability assessments. Using the first-order second-moment (FOSM) approach, they evaluated the influence of parameter uncertainty on embankment stability. Similarly, Gui et al. (2000) modelled slope stability under stochastic hydraulic conductivity conditions, integrating random field generation, seepage simulation, and limit equilibrium analysis.

Beyond estimation of soil parameters, uncertainty also affects the performance of geotechnical physical models. Popescu and Prevost (1995) emphasised that centrifuge and full-scale field tests—although valuable for studying soil-structure interaction—are subject to model scale effects, inconsistencies in sample preparation, instrumentation errors, and interpretation ambiguities. These factors introduce systematic and statistical errors into performance predictions.

Baecher and Christian (2003) provided a comprehensive discussion on the role of uncertainty in geotechnical decision-making, arguing for a balanced use of probabilistic and statistical tools within the bounds of practical engineering. Zhang et al. (2003; 2009) further demonstrated how different model assumptions—such as variations in hydrodynamic pressures or pullout resistance of soil nails—can lead to significantly different safety factors, reinforcing the importance of understanding model-related uncertainty. Moreover, empirical model validation has shown the limitations of conventional geotechnical approaches. Tang and Phoon (2021) conducted an extensive validation exercise using load test databases and observed that geotechnical models tend to be conservative and imprecise, which raises concerns about both safety and efficiency in design.

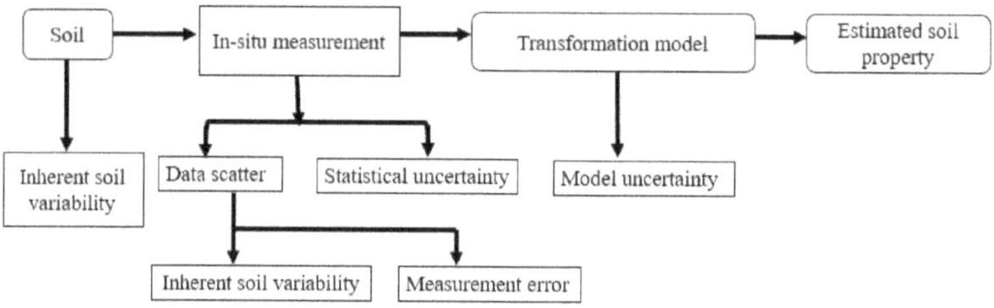

Fig 2.1. Source of uncertainty in geotechnical (Kulhawy 1992)

2.2 Bayesian Methods in Geotechnical Engineering: Theoretical Foundations and Applications

Bayesian methods, grounded in Bayes' theorem originally formulated by Thomas Bayes (Bayes & Price 1763), offer a probabilistic framework for statistical inference that is especially valuable in fields characterised by uncertainty and sparse data, such as geotechnical engineering. Unlike the frequentist perspective, which interprets probability as long-run frequency, the Bayesian viewpoint interprets probability as a measure of belief or confidence in a hypothesis, which can be updated as more evidence becomes available (Murphy 2012).

At the core of Bayesian inference lies the concept of updating prior beliefs about model parameters based on new data. This is mathematically expressed by the posterior distribution $p(\theta \mid D)$, which is derived using Bayes' theorem in equation 2.1.

$$p(\theta \mid D) = \frac{p(\theta) * p(D \mid \theta)}{p(D)} \tag{2.1}$$

where $p(\theta)$ represents prior knowledge, $p(D \mid \theta)$ is the likelihood of the observed data given the parameters, and $p(D)$ is the model evidence, which acts as a normalising constant (Murphy 2012). When analytical solutions are intractable, this evidence term is approximated using numerical methods like Markov Chain Monte Carlo (MCMC) (Jin et al., 2018), which enables sampling from complex posterior distributions. Popular MCMC algorithms include the Metropolis-Hastings method (Metropolis et al. 1953; Hastings 1970), Gibbs sampling (Geman & Geman 1984), and hybrid approaches (Gelman et al. 2014).

A key advantage of Bayesian models is their explicit quantification of uncertainty through credible intervals (CIs), which indicate the range within which a parameter value lies with a given level of credibility (Jin et al. 2018). When prior knowledge is vague or non-existent, non-informative priors can be employed to allow the data to primarily inform the posterior estimates (Wang & Cao 2013).

Bayesian modelling techniques are broadly categorized into regression for continuous outputs and classification for categorical outcomes (Johnson et al. 2021). Bayesian inference has gained increasing traction in geotechnical engineering due to its robustness in handling uncertainty, particularly in problems involving site characterization, reliability analysis, and the interpretation of subsurface properties. Wilson H. Tang's contributions exemplify the early integration of Bayesian reliability analysis in slope stability and offshore foundation design (Lacasse et al. 2017). One of the major advancements was linking Bayesian updating with the observational method first proposed by Peck (1969), enabling the assimilation of real-time data into predictive models (Baecher & Ladd 1997; Wu 2011; Papaioannou & Straub 2012; Spross & Larsson 2014; 2019).

Bayesian techniques, implemented via MCMC simulation, have been utilized in complex geotechnical applications such as deep excavation modeling (Juang et al. 2013b), embankments on soft soils (Zheng et al. 2018), and prediction of spudcan penetration resistance in layered soils (Uzielli et al. 2017; Li et al. 2018). Probabilistic back-analysis of slope failures using Bayesian inference has also seen widespread adoption (van der Krogt et al. 2021).

Traditionally, Bayesian methods have been instrumental in back-analyzing geotechnical parameters as random variables (Yang et al. 2018), and more recently, random fields using in-situ measurements. Such approaches support cumulative site knowledge updating and uncertainty quantification (Wang et al. 2016c; Cao et al. 2016). The integration of cone penetration test (CPT) data with surface wave analysis through Bayesian updating demonstrates its practical utility in subsurface characterization (Huang et al. 2018b).

The concept of within-site and cross-site variability in geotechnical datasets, initially highlighted by Baecher & Rackwitz (1982), has evolved to recognize group-level heterogeneity. Hierarchical Bayesian models have been proposed to utilize inter- and intra-group data for improved estimation accuracy (Bozorgzadeh & Bathurst 2020).

More recent advancements include non-parametric Bayesian approaches, such as Bayesian compressive sampling (Wang et al. 2022), and data-driven geotechnical interpolation models like the iterative convolution eXtreme Gradient Boosting (IC-XGBoost) model (Shi & Wang 2021). Additionally, Bayesian techniques have been used to refine design model uncertainties in offshore structures based on empirical data from hurricane events (Chen et al. 2020).

The influence of Bayesian thinking in geotechnical design continues to grow. Baecher (2017) emphasised its centrality in modern, data-driven engineering workflows. Gregory Baecher's 2021 Terzaghi Lecture further underscored the paradigm shift towards Bayesian-informed design and risk assessment. Bayesian methods offer a cohesive framework for integrating uncertainty, data, and prior knowledge in geotechnical engineering. Their ability to support both parameter estimation and decision-making under uncertainty has led to their adoption across a wide spectrum of problems, from site characterisation to advanced probabilistic modelling. As Phoon and Zhang (2022) highlight, Bayesian methods are now a cornerstone of modern machine learning applications in geotechnical systems.

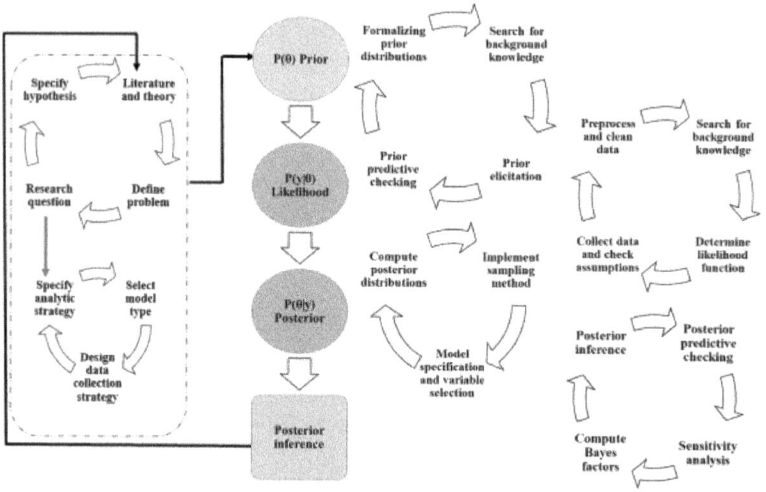

Fig 2.2. The cycle for Bayesian research (Rens et al.2021).

Table 2.1. Application of Bayesian methods in Geotechnical Engineering

Reference	Application	Description
Wang et al. (2012)	Settlement estimation	Estimated the maximum settlement caused by excavation activities
Houlsby and Houlsby (2013)	Soil strength profiling	Matched design profiles to undrained soil strength
Juang et al. (2013)	Back-analysis	Used back-analysis to determine soil parameters in braced excavations
Wang and Cao (2013)	Elastic modulus	Identified the undrained elastic modulus of clay soils
Zhou et al. (2013)	Shear stress prediction	Forecasted peak pullout shear stress in grouted soil nails
Ching and Phoon (2014)	Parameter variability	Quantified variability in clay soil parameters
Feng (2015)	Rock/tunnel properties	Predicted rock elastic modulus, rock mass deformation, and tunnel deformation
Miro et al. (2015)	Uncertainty impact	Assessed the impact of uncertain soil parameters on tunnel-related ground movement
Wang et al. (2015)	Site variability	Simulated variability of geotechnical parameters on a site-specific basis
Wang and Aladejare (2016)	Rock strength	Investigated the unconfined compressive strength and elastic modulus of rock
Wang and Akeju (2016)	Shear strength	Evaluated shear strength parameters, including cohesion and friction angle
Liu et al. (2016)	Soil modulus correlation	Explored correlations between resilient modulus and CPTu data in clay soils
Wang et al. (2016)	Site-specific properties	Defined site-specific soil property distributions
Ching et al. (2017)	Coarse soil parameters	Assessed geotechnical parameters of coarse-grained soils
Wang and Zhao (2017)	Design validation	Evaluated soil characteristics and validated the performance of design aids
Gong et al. (2017)	Site investigations	Enhanced efficiency of site investigations
Qi and Zhou (2017)	Parameter updating	Updated parameters for excavated sites using field data
Jin et al. (2018)	Braced excavation	Identified the most probable soil parameters for braced excavations
L. Wang et al. (2018)	soil-water characteristic curves	Determined soil-water characteristic curves based on site-specific data
X. Wang et al. (2018)	Subsurface prediction	Predicted subsurface stratigraphy and relevant geotechnical properties
Yang et al. (2019)	Granite modulus	Estimated the elastic modulus of granite using predictive methods
Bozorgzadeh et al. (2019)	Rock strength	Developed models to predict rock strength characteristics
Ching and Phoon (2019)	Multivariate modeling	Built multivariate probabilistic models for site-specific geotechnical data

Reference	Application	Description
Jin et al. (2019)	Soil model selection	Proposed a methodology for selecting advanced soil models under uncertainty
Hu et al. (2020)	Spatial estimation	Generated spatial estimates of geotechnical data from limited measurements
Hu and Wang (2020)	CPT classification	Incorporated uncertainty in CPT-based subsurface classification
Li et al. (2020)	Tunneling analysis	Applied probabilistic methods to analyse geomechanical parameters in tunnelling
Shi and Wang (2021a)	Stratigraphy estimation	Estimated subsurface stratigraphy from sparse geotechnical information

2.3. Bayesian Multilevel Modelling for Vs and N Relationship in Sandy Soil

2.3.1 Standard Penetration Test (SPT) blow count and Shear Wave Velocity (V_S) relationship

In geotechnical engineering, understanding subsurface soil properties is fundamental to designing safe and efficient structures. Two commonly employed methods for evaluating these properties are the Standard Penetration Test (SPT) and the measurement of Shear Wave Velocity (V_S) (Kramer 1996). The measured N-values are influenced by a range of factors, including soil type, water content, overburden pressure, hammer energy efficiency, and the method of drilling (Fabbrocino et al. 2015; Kirar et al. 2016; Rahimi et al. 2020).

Shear wave velocity (V_S) is another critical parameter for characterising the seismic behaviour of soil. It is frequently applied in studies related to stratigraphy, liquefaction potential, and site-specific seismic modelling (Daag et al. 2022). These approaches have been validated and accepted as reliable methods for capturing the dynamic response characteristics of subsurface materials (Garofalo et al. 2016a; 2016b; Foti et al. 2017; Agrawal and Cox 2019; Cox et al. 2019; Molnar et al. 2022).

Although direct field measurements of V_S are preferred, they are often expensive or logistically challenging, particularly in preliminary investigations or smaller projects. As a practical alternative, empirical correlations have been developed to estimate V_S from more readily available SPT-N values or Cone Penetration Test (CPT) data. These relationships are typically calibrated to local geological settings to improve their accuracy and applicability (Bajaj and Anbazhagan, 2019; Thokchom et al. 2017). The earliest developments of empirical V_S–N relationships can be traced to Japanese researchers in the 1970s who used regional datasets to analyse soil responses to earthquake-induced liquefaction (Imai and Yoshimura 1970; Shibata 1970; Ohta et al. 1978). These models were further refined by Imai and Tonouchi (1982), who introduced soil density, geologic age, and classification into the correlation frameworks. In the United States, Seed and Idriss (1981) and Seed et al. (1983) emphasised the importance of corrected N-values (N) to account for field variabilities when assessing liquefaction potential. Similarly, Brandenberg et al. (2010), through a study on California bridge sites, observed a significant dependency of both Vs and N on vertical effective stress.

Numerous empirical models have been proposed globally to capture V_s–N relationships, including those by Sykora and Stokoe (1983), Athanasopoulos (1994), Hasancebi and Ulusay (2007), Maheswari et al. (2010), and Anbazhagan et al. (2013). However, many of these correlations report relatively low coefficients of determination (R^2), indicating variability in predictive strength, often due to regional soil heterogeneities.

2.3.2 Hierarchical Modelling and Pooling Concepts in Geotechnical Data Analysis

In the analysis of grouped or structured data, Bayesian hierarchical modelling (BHM) has emerged as a powerful statistical approach, particularly suited to geotechnical engineering problems characterised by layered data from multiple sites. BHM allows for modelling parameters across different data groups under the assumption that while these parameters may not be identical, they are likely to be similar due to underlying geologic or geotechnical commonalities (Lunn et al. 2012). This framework is particularly advantageous for structured data where information can be "borrowed" across related groups, enabling improved parameter estimation even with sparse local observations (Bozorgzadeh et al. 2019).

A core idea within hierarchical modelling is the concept of pooling, which refers to how data from different groups are combined. There are three general strategies: no pooling, complete pooling, and partial pooling. In a no-pooling approach, each group is treated independently, assuming no relationship among them. Conversely, complete pooling aggregates all data into a single dataset, assuming homogeneity across groups and thus ignoring inter-group variability. Partial pooling, the middle ground, allows for both shared information and between-group variation, making it particularly useful in settings where groups are similar but not identical—a condition commonly encountered in geotechnical data (Gelman & Hill 2006).

Partial pooling is most often implemented via Bayesian hierarchical models, where group-specific parameters are considered exchangeable, drawn from a common prior distribution governed by higher-level (hyper)parameters. This enables a balance between group-specific estimation and global trend modelling (Gelman et al. 2013). The benefit lies in leveraging shared characteristics across sites while accommodating local site variability—an especially important trait in geotechnical contexts where data from individual sites may be sparse but geologic conditions can exhibit patterns of similarity.

Table 2.2. Application of Hierarchical Modelling in Geotechnical Engineering

Reference	Area	Application
Bozorgzadeh et al. (2019)	Rock Mechanics	Rock strength analysis using hierarchical modelling based on material classification
Phoon et al. (2019)	Data Integration	Introduced the concept of "Big Indirect Database" (BID) for integrating global geotechnical datasets
Ching & Phoon (2019, 2020)	Statistical Modelling	Multivariate probabilistic modelling to merge global trends with local data

Ching et al. (2021)	Statistical Modelling	Critique of complete pooling; proposed BHM integrating global trends and local site variability
Xiao et al. (2021)	Site Characterization	Empirical correlation between shear wave velocity (Vs) and SPT-N values
Feng et al. (2021a)	Rock Mechanics	In-situ stress in rock masses using BHM
Wu et al. (2022)	Statistical Modelling	Refined BHM to incorporate multimodal site distributions
Bozorgzadeh & Bathurst (2022)	Model Evaluation	Method bias evaluation in geotechnical models
Bozorgzadeh et al. (2023)	Foundation Engineering	Pile capacity prediction using BHM
Tao et al. (2023); Zhao et al. (2024)	Soil and Rock Characterisation	Soil and rock property estimation

2.4 Bayesian-Optimised Ensemble Learning for Bearing Capacity Prediction

2.4.1. Machine Learning and Artificial Intelligence Applications in Bearing Capacity Prediction

To overcome limitations of traditional methods, computational approaches, particularly Artificial Intelligence (AI), have gained traction in recent decades. AI models can capture complex nonlinear relationships between input variables such as soil properties and footing geometry, without requiring prior assumptions (Phoon & Zhang 2023; Zhang et al. 2021). Numerous applications of AI in geotechnical problems have been reported, including deep and shallow foundation design (Fatehnia & Amirinia, 2018; Mohammed et al. 2020). In terms of shallow foundations, Kalinli et al. (2011) developed an artificial neural network (ANN) model to predict UBC on granular soils and compared its performance with ANFIS. Padmini et al. (2008) applied soft computing tools like ANN, ANFIS, and FIS using five input features (B, D, L/B, γ, and φ), and concluded that ANFIS outperformed the others.

Further extending this line of work, Ahmad et al. (2021) implemented Gaussian Process Regression (GPR) to estimate the UBC of cohesionless soils, using input parameters including footing geometry and soil strength properties. Alzabeebee et al. (2023) proposed a multi-objective genetic algorithm model to enhance prediction accuracy and evaluated it using metrics like MAE, RMSE and R^2. Zhang and Xue (2022) highlighted the challenges in using standalone ANN models, including local minima issues and slow convergence. Similarly, Zhang et al. (2023) and Acharyya (2019) verified the efficacy of Finite Element Modelling (FEM) for analysing UBC, aligning well with experimental observations.

2.4.2. Ensemble Learning in Geotechnical Applications

As AI models diversified, ensemble learning emerged as a powerful strategy to improve prediction performance. By aggregating multiple learners, ensemble methods increase robustness and accuracy compared to individual models (Opitz & Maclin 1999). Ensemble models like Random Forest (RF), Gradient Boosted Trees (GBT), and Extra Trees (ETR) have shown promising results in geotechnical studies (Cheng & Lu 2020). For instance, Mangalathu et al. (2019) used RF to evaluate bridge damage states, while Zhang et al. (2020) applied ensemble techniques for interpolating missing values in rock burst datasets. Hou et al. (2022) implemented a superimposed ensemble classifier for rock classification using TBM data, outperforming standalone classifiers.

In slope stability prediction, Kardani et al. (2020) employed a hybrid stacking ensemble model, demonstrating enhanced accuracy. Lin et al. (2022) compared eight ensemble models for predicting slope safety factors and revealed significant performance disparities. Similarly, Arachchilage et al. (2023) used ensemble methods to predict the unconfined compressive strength of slag-based paste backfill. Zhang et al. (2021) demonstrated ensemble learning's success in predicting ground settlement due to excavations. These findings underscore the growing relevance and adaptability of ensemble models in solving complex geotechnical problems.

2.4.3. Bayesian Optimization for Hyperparameter Tuning in geotechnical engineering

A common challenge in AI and ensemble models is the selection of optimal hyperparameters, which significantly affect model performance. Bayesian Optimisation (BO) provides an efficient solution by treating the tuning task as a probabilistic optimisation problem. Acharya and Acharya (2019) investigated the reliability of shallow foundation bearing capacity on c-φ soil using FORM and SORM, with inputs like cohesion, unit weight, and friction angle. Although their focus was on reliability indices, the application of probabilistic techniques like BO can greatly enhance model calibration.

Bayesian optimisation has been integrated with ensemble methods in recent works to refine hyperparameters and improve predictive precision. Sun et al. (2022) proposed an ensemble-BO algorithm for slope stability prediction, demonstrating higher accuracy than traditional methods. Li (2021) employed BOA to fine-tune hyperparameters for rockburst prediction. Similarly, Bo et al. (2023) utilised BOA to determine optimal parameters in an ensemble classifier designed to assess tunnel squeezing risks. This growing trend suggests a shift towards integrating probabilistic and AI-based approaches to address uncertainty and nonlinearity in geotechnical design.

2.5. Soil Groutability and Hybrid Bayesian Network Approach

Permeation grouting is extensively used as a ground improvement technique to reduce the permeability of granular soils (Huang et al. 2007; Zebovitz et al. 1989). Among various grouting materials, microfine cement has gained prominence due to its superior ability to penetrate fine-grained soils and fill micro-cracks, along with the added benefit of environmental safety since it does not contaminate groundwater (Perret et al. 2000; Perret et al., 2002).

Groutability—the ability of a grout to effectively infiltrate a soil mass—is influenced by multiple soil and grout characteristics. While earlier predictive models largely relied on soil and grout particle size (Akbulut & Saglamer 2002), these models often prove inadequate for modern applications involving finer grouts. Particularly, their effectiveness diminishes with semi-nanometer scale particles, where grain size alone cannot reliably indicate groutability (Liao et al. 2011).

A number of researchers have highlighted the importance of considering additional soil and grout parameters. For instance, Akbulut and Saglamer (2002) and Ozgurel and Vipulanandan (2005) argued that parameters such as the water-to-cement ratio (w/c), void size within the soil, and fines content (FC) substantially affect grouting success. Liao et al. (2011) emphasised the role of soil gradation, including the coefficient of uniformity (C_u) and the coefficient of gradation (C_z), in enhancing prediction accuracy. These findings are corroborated by Tekin and Akbas (2011), who further added the importance of considering variables like relative density (D_r) and the $\frac{D_{15}}{D_{85}}$ ratio—a measure of the soil's gradation.

Other influential factors include grout viscosity and applied pressure. Meanwhile, grouting pressure (P) plays a pivotal role in expanding the treated zone but must exceed the grout's shear resistance for effective application (De Paoli et al.1992; Markou et al. 2018). Moreover, soil parameters such as relative density, fines content, and particle dispersion further affect groutability (Zebovitz et al. 1989; Karol 2003; Santagata & Santagata 2003; Jorne et al. 2015).

Artificial intelligence (AI) has emerged as a promising alternative to traditional formula-based approaches for predicting groutability, particularly under complex site-specific conditions. AI models are capable of learning from historical data and adapting to new information, making them well-suited for problems characterised by high uncertainty (Cheng et al. 2013). Since groutability prediction can be cast as a binary classification problem (i.e., success or failure), classification algorithms—particularly Artificial Neural Networks (ANNs)—have been widely explored. Studies by Chen et al. (2009); Kalinli et al. (2011); Liao et al. (2011), and Tekin and Akbas (2011) have demonstrated the feasibility of ANN models in this context. However, ANNs are not without limitations; their performance can be hindered by the need for extensive parameter tuning and a susceptibility to local minima during training (Bao et al. 2005; Kiranyaz et al. 2009). Furthermore, the "black box" nature of ANNs limits their interpretability, making them less desirable in decision-making scenarios involving engineering judgment.

Asadizadeh and Majdi (2019) developed adaptive neuro-fuzzy inference systems (ANFIS) to predict the groutability of granular soils. Their sensitivity analysis revealed that the most influential parameters were grain size indices (N_1 and N_2) and grouting pressure (P), while water-cement ratio, fines content, and relative density were found to be less impactful. Similarly, Tekin and Akbas (2011) ranked the predictive importance of $\frac{D_{15}}{D_{85}}$, pressure, w/c ratio, relative density, and fines content, reinforcing the need for a multi-variable modelling approach. Collectively, these studies underscore the multifactorial nature of groutability and point toward the need for robust, data-driven models that integrate soil, grout, and site-specific parameters. While AI-based models like ANNs and ANFIS offer significant advantages, the unexplored potential of Bayesian classifiers presents an exciting avenue for research in soil Groutability prediction.

2.5.1 Bayesian Networks and Geotechnical Engineering

Bayesian Networks (BNs), introduced by Pearl (1988), are powerful tools for probabilistic reasoning under uncertainty. They belong to the broader class of probabilistic graphical models and are structured as directed acyclic graphs (DAGs), where nodes represent variables and edges represent conditional dependencies. This structure enables BNs to model complex, nonlinear systems, making them particularly useful for tasks involving inference, classification, regression, and system characterisation (Larrañaga & Moral 2011).

One of the key strengths of BNs lies in their solid foundation in probability theory and their capability to incorporate domain knowledge, prior beliefs, and heterogeneous data sources into a unified analytical framework (Hu et al. 2016). Unlike conventional machine learning techniques such as artificial neural networks (ANNs) or linear regression (LR), BNs provide several advantages:

1. They support inference even with incomplete data;
2. They enable both forward (predictive) and backwards (diagnostic) reasoning;
3. They are less prone to overfitting;
4. They offer transparent decision-support systems, with interpretable graphical representations;
5. They allow for continuous updating of probabilities as new data becomes available (Liu & Hu 2020).

In geotechnical engineering, BNs have been explored for predicting soil behaviour under seismic loading. Hu and Liu (2019a;2019b) developed BN-based models to predict the liquefaction potential of various soil types. However, these studies included a limited number of gravel soil samples (only 50 cases), which could affect model generalizability. Zhang et al. (2014) utilised fuzzy Bayesian networks (FBNs) to investigate the causal interactions between tunnel-induced damage and its contributing factors. Chung et al. (2019) focused on identifying and characterising potential risk events that could arise throughout the tunnelling process, enhancing proactive risk mitigation strategies. Zhou et al. (2018) introduced a BN-driven risk analysis model specifically aimed at assessing the safety of diaphragm wall deflection during deep excavations. This probabilistic approach enabled better prediction and interpretation of structural responses under uncertain conditions. J. Zhang et al. (2020) applied Bayesian principles for the calibration of resistance factors in pile foundation design, providing a reliability-based design framework that accounts for geotechnical uncertainty and variability. Wang and Chen (2017) developed an integrated risk assessment approach by coupling Bayesian networks with the fuzzy comprehensive evaluation method (FCEM) to manage uncertainties in metro infrastructure projects.

2.5.2 Use of Bayesian networks in hybrid datasets

Aguilera et al. (2011) observed that most BN applications in environmental sciences rely on discrete or discretised data, despite the prevalence of continuous or hybrid datasets. This limitation often results in information loss, as continuous variables are discretised to fit traditional BN frameworks (Uusitalo 2007). To address this, several extensions have been proposed that permit the use of continuous data within BN structures. These include

Mixtures of Polynomials (MoP) (Shenoy & West 2011), Mixtures of Truncated Basis Functions (MoTBFs) (Langseth et al. 2012. Given that groutability is influenced by both continuous variables (e.g., particle size, pressure) and discrete outcomes (e.g., success/failure), hybrid BNs present a promising framework.

2.6. Probabilistic Approach for Prediction of Liquefaction

Soil liquefaction a process where saturated, cohesionless soils lose strength and stiffness due to earthquake-induced cyclic loading is among the most destructive consequences of seismic events. This phenomenon has contributed significantly to infrastructure damage during major earthquakes, including the 1964 Niigata (Hamada et al. 1987; Ishihara et al. 1981; 1995 Hanshin (Hatanaka et al. 1997), 1999 Chi-Chi (Yuan et al. 2004), 2008 Wenchuan (Cao et al. 2011; Chen et al. 2018), 2010 Maule (Bertalot et al. 2013), and 2018 Palu earthquakes (Sassa et al. 2019). Given its catastrophic consequences, predicting liquefaction has been a central focus in geotechnical engineering for decades.

2.6.1 Conventional Approaches

Early research efforts relied heavily on laboratory-based undrained cyclic loading tests to evaluate liquefaction susceptibility (Castro 1975; Peck 1979; Castro 1987). However, due to the challenge of retrieving undisturbed sandy soil samples, in situ testing gained popularity. Methods such as the Standard Penetration Test (SPT), Cone Penetration Test (CPT), Shear Wave Velocity (V_s), and Boring Pressuremeter Test (BPT) have formed the empirical foundation for liquefaction evaluation (Seed et al. 1983; Juang et al. 2014; Kayen & Mitchell 1997).

Empirical approaches typically consist of two main components: (i) identifying a field-measurable index reflective of liquefaction potential, and (ii) building models grounded in historical earthquake performance data. Techniques like the Seed-Idriss simplified method (Seed & Idriss 1971), V_s-based methods (Andrus and Stokoe 2000; Kayen et al. 2013), and static or dynamic laboratory tests (Yu et al. 2012; Hu et al. 2017) have contributed to this approach. Despite their usefulness, these methods are often costly and time-consuming due to the reliance on extensive field and lab investigations.

2.6.2 Emergence of Machine Learning-Based Models

To overcome the limitations of conventional techniques, researchers have increasingly adopted machine learning (ML) methods, which offer scalability, speed, and the ability to capture complex nonlinear relationships (Karthikeyan & Samui 2014; Ghani & Kumari 2022). These data-driven models leverage large datasets derived from field observations and earthquake case histories.

For instance, Tung et al. (1993) introduced an artificial neural network (ANN) model trained on historical earthquake data to forecast liquefaction events. Juang et al. (2001) were among the early researchers to integrate both logistic regression and Bayesian techniques for this purpose. They developed a predictive formula for estimating the probability of liquefaction based on Vs, emphasising the advantages of probabilistic approaches in capturing uncertainty in geotechnical data. Kumar et al. (2021) proposed a deep learning model using emotionally modulated backpropagation (EMBP) for soil classification. Similarly, Ahmad et al. (2021) assessed the predictive capability of multiple ML algorithms using CPT-derived parameters. Garcia et al. (2012) developed a hybrid ML framework integrating geotechnical and seismic variables to evaluate liquefaction potential. Kayen et al. (2013) adopted a Bayesian updating framework to refine the assessment of liquefaction potential. Their model leveraged classification trees and neural networks trained on databases containing CPT and Vs values. Overall, ML models have demonstrated strong performance in classifying soils as liquefied or non-liquefied (Chen et al. 2018). Shen et al. (2016) also formulated a liquefaction probability model using logistic regression, reinforcing the practicality of regression-based methods for empirical evaluation. Machine learning and probabilistic methods have become increasingly prominent in recent liquefaction studies, with researchers applying a wide range of algorithms and hybrid approaches. A notable stream of work has focused on ensemble learning. Decision trees, random forests, gradient boosting, XGBoost, AdaBoost, and Extra Trees have attracted considerable interest across engineering applications (Tiwari et al. 2024; Jas and Dodagoudar 2023b; Chen et al. 2018), although their application to liquefaction problems is still comparatively limited. Most existing studies test individual ensemble models rather than evaluating several ensemble families together, leaving an opportunity for broader comparative assessments. Several authors have examined liquefaction potential using SPT- and Vs-based databases. Feng et al. (2024) demonstrated that a CatBoost model yields higher predictive accuracy than various other ML techniques for SPT-based classification. Using SPT datasets, Kurnaz et al. (2023) reported that a hyperparameter-tuned ANN performed better than the other evaluated models. Sui et al. (2023) approached the problem using ten ML algorithms with Vs data. Studies using historical datasets have also been conducted; for example, Demir et al. (2024) tested multiple ensemble learners and identified LightGBM as the strongest performer while emphasizing the importance of choosing soil-type-consistent input variables. CPT-based evaluations show similar trends. Khatti et al. (2024a) assessed the influence of varying numbers of trees and classifiers, and Demir and Sahin (2023) explored multiple forest-type models including RF, canonical correlation forest, and rotation forest. Parallel to deterministic ML approaches, probabilistic and hybrid frameworks have also been widely explored. Bayesian belief networks, probabilistic neural networks, response surface methods, evolutionary approaches, and fuzzy logic have all been used to assess liquefaction potential using both laboratory and in-situ data (Jas and Dodagoudar 2023). Hybrid numerical–probabilistic algorithms are also gaining traction, particularly for improving classification reliability (Tiwari et al. 2024; Jas and Dodagoudar 2023).

Liquefaction datasets are often imbalanced, which can lead ML models to favour majority classes. To address this issue, several studies recommend ensemble algorithms that re-weight difficult cases, alternative metrics such as G-mean (Tiwari et al. 2024; Das et al. 2020), and stratified cross-validation for more reliable model assessment. Overfitting and underfitting remain persistent challenges, but ensemble techniques—through diverse base learners and varied data subsets—tend to offer better generalization and reduce the likelihood of models settling into local minima (Jas and Dodagoudar 2023b). Another major limitation is the weak physical interpretability of conventional AI/ML models. To bridge this gap, recent work has explored physics-informed networks, hybrid transfer-learning approaches, and collocation techniques, all aimed at embedding mechanical behaviour or governing relationships directly into the learning process (Tiwari et al. 2024; Jas et al. 2025). Deep learning has also been used to model dynamic responses such as excess pore pressure and shear strain (Jas et al. 2024), with LSTM and convolutional models forming the basis for interpretable hybrid frameworks (Ghani et al. 2025). Model transparency has received particular attention through explainable AI (XAI). SHAP, LIME, partial dependence plots, and cohort-based techniques have been introduced to reveal how ML models behave at both global and local scales. Jas and Dodagoudar (2023) combined SHAP with XGBoost for CPT-based liquefaction assessment, and Jas et al. (2024) developed LightGBM-SHAP models for evaluating gravelly soils. Such tools help clarify feature roles, which is especially important since conventional sensitivity analysis often depends on the chosen model and its hyperparameters (Das et al. 2020). Looking forward, physics-informed and XAI-supported approaches show strong promise for developing data-driven constitutive insights and improving the reliability of liquefaction prediction under dynamic loading (Jas et al. 2025).

2.7 Research Gap

Despite significant advances in empirical modelling and the adoption of machine learning techniques within geotechnical engineering, there remains a persistent gap in the development of generalizable, interpretable, and uncertainty-aware predictive models. Current methodologies ranging from deterministic models to black-box ML approaches, frequently neglect the probabilistic nature of subsurface behaviour and are often based on region-specific datasets, which limits their applicability across diverse geological contexts.

For instance, the widely studied correlation between Standard Penetration Test (SPT) blow counts and shear wave velocity (Vs) in sandy soils often fails to accommodate inter-regional variability and local geological deviations. While some probabilistic frameworks have emerged, they rarely leverage the hierarchical structure of site investigation data. The underutilization of Bayesian multilevel modelling, which can simultaneously capture global patterns and site-specific variations, represents a critical gap, particularly for applications such as seismic site response analysis and regional hazard assessments.

Similarly, ensemble learning methods have been explored for bearing capacity prediction, but these models frequently lack rigorous hyperparameter optimisation and are susceptible to overfitting, especially in heterogeneous geotechnical datasets. Although Bayesian Optimisation provides a principled and data-efficient solution for tuning model parameters under uncertainty, it remains underexplored in geotechnical contexts, particularly for transferring ensemble models across different geotechnical settings.

In the domain of groutability prediction, empirical indices and conventional machine learning techniques have proven insufficient in representing the complex, nonlinear interactions among soil gradation, grout properties, and injection conditions. Hybrid Bayesian Networks capable of integrating continuous and discrete variables offer a promising alternative, yet their application in this domain is in its infancy. Furthermore, issues such as class imbalance and sampling bias remain largely unaddressed, reducing model reliability and interpretability.

Another underdeveloped area is the probabilistic modelling of soil liquefaction potential. Existing models, including those based on simplified procedures or CPT-derived empirical equations, often lack statistical rigour and fail to validate critical assumptions underlying logistic regression analysis. Most studies do not examine whether explanatory variables conform to basic statistical criteria (e.g., normality, independence, and linearity with the logit function), raising concerns about the reliability and transferability of their findings. Bayesian logistic regression presents a compelling alternative offering transparency, statistical rigour, but its use remains limited, especially in integrating diverse datasets across seismic regions.

Taken together, these gaps underscore the need for a unified Bayesian modelling framework in geotechnical engineering, one that not only accounts for uncertainty and data heterogeneity but also ensures interpretability and transferability. By embedding geotechnical domain knowledge within a probabilistic structure, Bayesian methods can bridge the divide between empirical practicality and statistical robustness. This thesis aims to advance such a framework through the application of Bayesian multilevel modelling, Bayesian-optimised ensemble methods, hybrid Bayesian networks, and Bayesian logistic regression, contributing to more reliable and generalizable models for Vs estimation, bearing pressure prediction, groutability assessment, and liquefaction hazard evaluation.

2.8 Aim and Objectives

The aim of this study is to employ the Bayesian framework for effectively managing geotechnical data across multiple domains. Specifically, it focuses on its application in hierarchical modelling, where uncertainties in geotechnical parameters are systematically quantified; optimisation, where Bayesian techniques aid in parameter estimation and model refinement; and classification, where probabilistic approaches enhance data-driven decision-making. Through these applications, the study aims to improve the reliability and accuracy of geotechnical predictions while addressing inherent uncertainties in the data.

The objectives of the study are

1. To develop a Bayesian multilevel modelling framework for the calibration of shear wave velocity (Vs) using Standard Penetration Test (SPT) blow counts in sandy soils, accounting for site-specific variability and measurement uncertainty.

2. To enhance the predictive accuracy of ensemble learning models for estimating the bearing pressure of shallow footings on clayey soils by integrating Bayesian optimisation techniques for hyperparameter tuning and model generalisation.

3. To construct a hybrid Bayesian Network model for predicting the groutability of granular soils, capturing complex interdependencies among geotechnical variables.

4. To implement a Bayesian logistic regression model for probabilistic prediction of soil liquefaction using CPT data, for improved reliability in the classification of liquefaction cases.

2.9 Scope of the Study

This research explores the integration of Bayesian methods in geotechnical engineering by addressing four distinct, yet interconnected, problem domains: empirical correlation modelling, bearing capacity prediction, soil groutability assessment, and liquefaction susceptibility classification. The overarching scope encompasses the formulation, implementation, and validation of Bayesian statistical frameworks, both parametric and graphical, across different geotechnical applications. The scope of each objective in given below

1. Bayesian Multilevel Modelling for Vs–SPT Correlation

The study formulates a Bayesian hierarchical regression model to quantify and generalise the relationship between Standard Penetration Test (SPT) blow count (N) and Shear Wave Velocity (Vs) in sandy soils. By allowing intercepts to vary by geographical region while constraining slope variability, the model captures both intra- and inter-regional heterogeneity. The scope includes model comparison via complete pooling, no pooling, and partial pooling approaches; transformation of empirical N–Vs datasets to a log-linear power-law form; and performance evaluation under unseen or new data scenarios.

2. Bayesian-Optimised Ensemble Learning for Bearing Capacity Prediction

The scope extends to the development of hybrid ensemble learning models (AdaBoost-BOA, GBoost-BOA) trained on literature-compiled Plate Load Test (PLT) data. Bayesian Optimisation is employed to systematically tune hyperparameters, enhancing generalisation and minimising overfitting risks. The research benchmarks these models against conventional ANNs, focusing on accuracy, robustness, and applicability to shallow foundation systems on cohesive soils.

3. Hybrid Bayesian Network Model for Groutability Assessment

A data-driven hybrid Bayesian Network (BN) is constructed to predict grouting success in coarse-grained soils treated with microfine cement. The model integrates continuous and categorical variables without discretisation, overcoming limitations of classical BNs. The model is designed to aid engineering decision-making by enabling transparent probabilistic inference.

4. Bayesian Logistic Regression for Liquefaction Potential Prediction

The final module involves the development of a Bayesian logistic regression (BLR) model to probabilistically classify soil liquefaction based on seismic and geotechnical indicators. Various transformative models for various soil types are explored to classify samples as liquefied or non-liquefied. The performance of the model is compared with empirical approaches.

Overall Boundaries and Assumptions

1. The models are limited to field-scale secondary datasets sourced from published literature.

2. Prior distributions in Bayesian models are either weakly informative or empirically derived, based on data availability.
3. Emphasis is placed on statistical modelling and algorithmic development, with no physical experimentation.

The scope of this thesis lies in demonstrating that Bayesian methods provide a coherent and interpretable framework for geotechnical problems traditionally dominated by empirical and black-box models. By embedding domain knowledge through priors, leveraging probabilistic inference, and using modern computational tools like MCMC and Bayesian Optimisation, the study bridges the gap between statistical rigour and engineering applicability. The findings aim to guide practitioners toward more reliable, uncertainty-aware decision-making in seismic hazard analysis, foundation design, and soil improvement techniques.

Reference

Acharya, Mahesh, and Indra Prasad Acharya. "Reliability Analysis of Bearing Capacity of Shallow Foundation on C- φ Soil." Journal of Advanced College of Engineering and Management, vol. 5, Dec. 2019, pp. 71–78. Crossref, https://doi.org/10.3126/jacem.v5i0.26690.

Acharyya, R. "Finite Element Investigation and ANN-Based Prediction of the Bearing Capacity of Strip Footings Resting on Sloping Ground." International Journal of Geo-Engineering, vol. 10, no. 1, Mar. 2019. Crossref, https://doi.org/10.1186/s40703-019-0100-z.

Agrawal, Nirupama, and Robin Cox. "Natural Disasters and 150th Commemoration of Canada as a Country." Natural Hazards, vol. 98, no. 1, Aug. 2019, pp. 3–7. Crossref, https://doi.org/10.1007/s11069-019-03751-9.

Aguilera, P. A., et al. "Bayesian Networks in Environmental Modelling." Environmental Modelling & Software, vol. 26, no. 12, Dec. 2011, pp. 1376–88. Crossref, https://doi.org/10.1016/j.envsoft.2011.06.004.

Ahmad, Mahmood, et al. "Evaluation of Liquefaction-Induced Lateral Displacement Using Bayesian Belief Networks." Frontiers of Structural and Civil Engineering, vol. 15, no. 1, Feb. 2021, pp. 80–98. Crossref, https://doi.org/10.1007/s11709-021-0682-3.

Ahmad, Mahmood, et al. "Prediction of Ultimate Bearing Capacity of Shallow Foundations on Cohesionless Soils: A Gaussian Process Regression Approach." Applied Sciences, vol. 11, no. 21, Nov. 2021, p. 10317. Crossref, https://doi.org/10.3390/app112110317.

Akbulut, Suat, and Ahmet Saglamer. "Estimating the Groutability of Granular Soils: A New Approach." Tunnelling and Underground Space Technology, vol. 17, no. 4, Oct. 2002, pp. 371–80. Crossref, https://doi.org/10.1016/s0886-7798(02)00040-8.

Alzabeebee, Saif, et al. "New Model to Predict Bearing Capacity of Shallow Foundations Resting on Cohesionless Soil." Geotechnical and Geological Engineering, vol. 41, no. 6, May 2023, pp. 3531–47. Crossref, https://doi.org/10.1007/s10706-023-02472-y.

Ames, Daniel P., et al. "Using Bayesian Networks to Model Watershed Management Decisions: An East Canyon Creek Case Study." Journal of Hydroinformatics, vol. 7, no. 4, Oct. 2005, pp. 267–82. Crossref, https://doi.org/10.2166/hydro.2005.0023.

Anbazhagan, P., et al. "Seismic Site Classification and Correlation between Standard Penetration Test N Value and Shear Wave Velocity for Lucknow City in Indo-Gangetic Basin." Pure and Applied Geophysics, vol. 170, no. 3, July 2012, pp. 299–318. Crossref, https://doi.org/10.1007/s00024-012-0525-1.

Andrus, Ronald D., et al. Draft Guidelines for Evaluating Liquefaction Resistance Using Shear Wave Velocity Measurements and Simplified Procedures. National Institute of Standards and Technology, 1999. Crossref, https://doi.org/10.6028/nist.ir.6277.

Asadizadeh, Mostafa, and Abbas Majdi. "Developing new Adaptive Neuro-Fuzzy Inference System models to predict granular soil groutability." International Journal of Mining and Geo-Engineering 53.2 (2019): 133-142.

Athanasopoulos, G. A. "Empirical correlations v sub so-n sub spt for soils of Greece: A comparative study of reliability." Seventh International Conference on Soil Dynamics and Earthquake Engineering (SDEE 95) Hellenic Society of Soil Mechanics and Foundation Engineering, Greece.. 1995.

Baecher, Gregory B. "Bayesian Thinking in Geotechnics." Geo-Risk 2017, June 2017, pp. 1–18. Crossref, https://doi.org/10.1061/9780784480694.001.

Baecher, Gregory B., and Charles C. Ladd. "Formal Observational Approach to Staged Loading." Transportation Research Record: Journal of the Transportation Research Board, vol. 1582, no. 1, Jan. 1997, pp. 49–52. Crossref, https://doi.org/10.3141/1582-08.

Baecher, Gregory B., and John T. Christian. Reliability and statistics in geotechnical engineering. John Wiley & Sons, 2005.

Baecher, Gregory B., and Rüdiger Rackwitz. "Factors of Safety and Pile Load Tests." International Journal for Numerical and Analytical Methods in Geomechanics, vol. 6, no. 4, Oct. 1982, pp. 409–24. Portico, Crossref, https://doi.org/10.1002/nag.1610060404.

Bajaj, Ketan, and P. Anbazhagan. "Seismic Site Classification and Correlation between VS and SPT-N for Deep Soil Sites in Indo-Gangetic Basin." Journal of Applied Geophysics, vol. 163, Apr. 2019, pp. 55–72. Crossref, https://doi.org/10.1016/j.jappgeo.2019.02.011.

Balasooriya Arachchilage, Chathuranga, et al. "Forecasting Unconfined Compressive Strength of Calcium Sulfoaluminate Cement Mixtures Using Ensemble Machine Learning Techniques Integrated with Shapely-Additive Explanations." Construction and Building Materials, vol. 409, Dec. 2023, p. 134083. Crossref, https://doi.org/10.1016/j.conbuildmat.2023.134083.

Barbour, S. Lee, and John Krahn. "Numerical modelling–prediction or process." *Geotechnical News* 22.4 (2004): 44-52.

Bayes, Thomas. "An essay towards solving a problem in the doctrine of chances." Biometrika 45.3-4 (1958): 296-315.

BERTALOT, D., et al. "Influence of Bearing Pressure on Liquefaction-Induced Settlement of Shallow Foundations." Géotechnique, vol. 63, no. 5, Apr. 2013, pp. 391–99. Crossref, https://doi.org/10.1680/geot.11.p.040.

Bo, Yin, et al. "Robust Model for Tunnel Squeezing Using Bayesian Optimized Classifiers with Partially Missing Database." Underground Space, vol. 10, June 2023, pp. 91–117. Crossref, https://doi.org/10.1016/j.undsp.2022.11.001.

Bozorgzadeh, Nezam, and Richard J. Bathurst. "Hierarchical Bayesian Approaches to Statistical Modelling of Geotechnical Data." Georisk: Assessment and Management of Risk for Engineered Systems and Geohazards, vol. 16, no. 3, Dec. 2020, pp. 452–69. Crossref, https://doi.org/10.1080/17499518.2020.1864411.

Bozorgzadeh, Nezam, et al. "Hierarchical Bayesian Modelling of Geotechnical Data: Application to Rock Strength." Géotechnique, vol. 69, no. 12, Dec. 2019, pp. 1056–70. Crossref, https://doi.org/10.1680/jgeot.17.p.282.

Bozorgzadeh, Nezam, et al. "Model Calibration: A Hierarchical Bayesian Approach." Probabilistic Engineering Mechanics, vol. 71, Jan. 2023, p. 103379. Crossref, https://doi.org/10.1016/j.probengmech.2022.103379.

Brandenberg, Scott J., et al. "Shear Wave Velocity as Function of Standard Penetration Test Resistance and Vertical Effective Stress at California Bridge Sites." Soil Dynamics and Earthquake Engineering, vol. 30, no. 10, Oct. 2010, pp. 1026–35. Crossref, https://doi.org/10.1016/j.soildyn.2010.04.014.

Cao, Zhenzhong, et al. "Gravelly Soils That Liquefied during 2008 Wenchuan, China Earthquake, Ms=8.0." Soil Dynamics and Earthquake Engineering, vol. 31, no. 8, Aug. 2011, pp. 1132–43. Crossref, https://doi.org/10.1016/j.soildyn.2011.04.001.

Cao, Zi-Jun, et al. "Site-Specific Characterization of Soil Properties Using Multiple Measurements from Different Test Procedures at Different Locations – A Bayesian Sequential Updating Approach." Engineering Geology, vol. 211, Aug. 2016, pp. 150–61. Crossref, https://doi.org/10.1016/j.enggeo.2016.06.021.

Castro, G. "On the Behavior of Soils During Earthquakes – Liquefaction." Soil Dynamics and Liquefaction, 1987, pp. 169–204. Crossref, https://doi.org/10.1016/b978-0-444-98958-1.50017-0.

Castro, Gonzalo. "Liquefaction and Cyclic Mobility of Saturated Sands." Journal of the Geotechnical Engineering Division, vol. 101, no. 6, June 1975, pp. 551–69. Crossref, https://doi.org/10.1061/ajgeb6.0000173.

Chen, Caroline J., and C. Hsein Juang. "Calibration of SPT- and CPT-Based Liquefaction Evaluation Methods." Innovations and Applications in Geotechnical Site Characterisation, July 2000, pp. 49–64. Crossref, https://doi.org/10.1061/40505(285)4.

Chen, Jinbo, et al. "Calibration of Model Uncertainties for Fixed Steel Offshore Platforms Based on Observed Performance in Gulf of Mexico Hurricanes." Journal of Geotechnical and Geoenvironmental Engineering, vol. 146, no. 6, June 2020. Crossref, https://doi.org/10.1061/(asce)gt.1943-5606.0002247.

Chen, Weiyun, et al. "Momentary Liquefaction of Porous Seabed under Vertical Seismic Action." Applied Ocean Research, vol. 73, Apr. 2018, pp. 80–87. Crossref, https://doi.org/10.1016/j.apor.2018.02.005.

Chen, Yong-gui, et al. "Strength of Copolymer Grouting Material Based on Orthogonal Experiment." Journal of Central South University of Technology, vol. 16, no. 1, Feb. 2009, pp. 143–48. Crossref, https://doi.org/10.1007/s11771-009-0024-4.

Cheng, Kai, and Zhenzhou Lu. "Structural Reliability Analysis Based on Ensemble Learning of Surrogate Models." Structural Safety, vol. 83, Mar. 2020, p. 101905. Crossref, https://doi.org/10.1016/j.strusafe.2019.101905.

Cheng, Y. M., et al. "An Innovative Geonail System for Soft Ground Stabilization." Soils and Foundations, vol. 53, no. 2, Apr. 2013, pp. 282–98. Crossref, https://doi.org/10.1016/j.sandf.2013.02.009.

Ching, Jianye, and Kok-Kwang Phoon. "Constructing a Site-Specific Multivariate Probability Distribution Using Sparse, Incomplete, and Spatially Variable (MUSIC-X) Data." Journal of Engineering Mechanics, vol. 146, no. 7, July 2020. Crossref, https://doi.org/10.1061/(asce)em.1943-7889.0001779.

Ching, Jianye, and Kok-Kwang Phoon. "Constructing Site-Specific Multivariate Probability Distribution Model Using Bayesian Machine Learning." Journal of Engineering Mechanics, vol. 145, no. 1, Jan. 2019. Crossref, https://doi.org/10.1061/(asce)em.1943-7889.0001537.

Ching, Jianye, and Kok-Kwang Phoon. "Correlations among Some Clay Parameters — the Multivariate Distribution." Canadian Geotechnical Journal, vol. 51, no. 6, June 2014, pp. 686–704. Crossref, https://doi.org/10.1139/cgj-2013-0353.

Ching, Jianye, et al. "Constructing Quasi-Site-Specific Multivariate Probability Distribution Using Hierarchical Bayesian Model." Journal of Engineering Mechanics, vol. 147, no. 10, Oct. 2021. Crossref, https://doi.org/10.1061/(asce)em.1943-7889.0001964.

Ching, Jianye, et al. "Correlations among Some Parameters of Coarse-Grained Soils — the Multivariate Probability Distribution Model." Canadian Geotechnical Journal, vol. 54, no. 9, Sept. 2017, pp. 1203–20. Crossref, https://doi.org/10.1139/cgj-2016-0571.

Ching, Jianye, et al. "Quasi-Site-Specific Prediction for Deformation Modulus of Rock Mass." Canadian Geotechnical Journal, vol. 58, no. 7, July 2021, pp. 936–51. Crossref, https://doi.org/10.1139/cgj-2020-0168.

Christian, John T., et al. "Reliability Applied to Slope Stability Analysis." Journal of Geotechnical Engineering, vol. 120, no. 12, Dec. 1994, pp. 2180–207. Crossref, https://doi.org/10.1061/(asce)0733-9410(1994)120:12(2180).

Chung, Heeyoung, et al. "Bayesian Networks-Based Shield TBM Risk Management System: Methodology Development and Application." KSCE Journal of Civil Engineering, vol. 23, no. 1, Jan. 2019, pp. 452–65. Crossref, https://doi.org/10.1007/s12205-018-0912-y.

Cox, Brady R., et al. "A Statistical Representation and Frequency-Domain Window-Rejection Algorithm for Single-Station HVSR Measurements." Geophysical Journal International, vol. 221, no. 3, Mar. 2020, pp. 2170–83. Crossref, https://doi.org/10.1093/gji/ggaa119.

Curran, J. H., and R. E. Hammah. "Keynote lecture: Seven lessons of geomechanics software development." Proceedings of the 41st US Rock Mechanics Symposium, Golden Rocks 2006, 50 Years of Rock Mechanics, June 17-21. 2006.

Daag, Arturo S., et al. "Empirical Correlation between Standard Penetration Resistance (SPT-N) and Shear Wave Velocity (Vs) for Soils in Metro Manila, Philippines." Applied Sciences, vol. 12, no. 16, Aug. 2022, p. 8067. Crossref, https://doi.org/10.3390/app12168067.

Das, Sarat Kumar, et al. "Multi-Objective Feature Selection (MOFS) Algorithms for Prediction of Liquefaction Susceptibility of Soil Based on in Situ Test Methods." Natural Hazards, vol. 103, no. 2, June 2020, pp. 2371–93. Crossref, https://doi.org/10.1007/s11069-020-04089-3.

De Paoli, B., et al. "Fundamental observations on cement based grouts (1): traditional materials." Proc. of Conf. on Grouting in Geotechnical Engineering, New Orleans, ASCE, New York. 1992.

Demir, Alparslan Serhat, et al. "A Comparative Analysis of Ensemble Learning Algorithms with Hyperparameter Optimization for Soil Liquefaction Prediction." Environmental Earth Sciences, vol. 83, no. 9, May 2024. Crossref, https://doi.org/10.1007/s12665-024-11600-7.

earl, Judea. "Evidential Reasoning Under Uncertainty." Exploring Artificial Intelligence, 1988, pp. 381–418. Crossref, https://doi.org/10.1016/b978-0-934613-67-5.50014-9.

Fabbrocino, Silvia, et al. "SPT Blow Count vs. Shear Wave Velocity Relationship in the Structurally Complex Formations of the Molise Region (Italy)." Engineering Geology, vol. 187, Mar. 2015, pp. 84–97. Crossref, https://doi.org/10.1016/j.enggeo.2014.12.016.

Fang, Yu, et al. "Neural Transfer Learning for Soil Liquefaction Tests." Computers & Geosciences, vol. 171, Feb. 2023, p. 105282. Crossref, https://doi.org/10.1016/j.cageo.2022.105282.

Fatehnia, Milad, and Gholamreza Amirinia. "A Review of Genetic Programming and Artificial Neural Network Applications in Pile Foundations." International Journal of Geo-Engineering, vol. 9, no. 1, Jan. 2018. Crossref, https://doi.org/10.1186/s40703-017-0067-6.

Feng, Xianda. Application of Bayesian Approach in Geotechnical Engineering. Crossref, https://doi.org/10.20868/upm.thesis.37270.

Feng, Yu, et al. "Improving Local Mean Stress Estimation Using Bayesian Hierarchical Modelling." International Journal of Rock Mechanics and Mining Sciences, vol. 148, Dec. 2021, p. 104924. Crossref, https://doi.org/10.1016/j.ijrmms.2021.104924.

Foti, Sebastiano, et al. "Guidelines for the Good Practice of Surface Wave Analysis: A Product of the InterPACIFIC Project." Bulletin of Earthquake Engineering, vol. 16, no. 6, Sept. 2017, pp. 2367–420. Crossref, https://doi.org/10.1007/s10518-017-0206-7.

García, S., et al. "Liquefaction assessment through machine learning approach." 15th World Conference on Earthquake Engineering. 2012.

Garofalo, F., et al. "InterPACIFIC Project: Comparison of Invasive and Non-Invasive Methods for Seismic Site Characterization. Part I: Intra-Comparison of Surface Wave Methods." Soil Dynamics and Earthquake Engineering, vol. 82, Mar. 2016, pp. 222–40. Crossref, https://doi.org/10.1016/j.soildyn.2015.12.010.

Garofalo, F., et al. "InterPACIFIC Project: Comparison of Invasive and Non-Invasive Methods for Seismic Site Characterization. Part II: Inter-Comparison between Surface-Wave and Borehole Methods." Soil Dynamics and Earthquake Engineering, vol. 82, Mar. 2016, pp. 241–54. Crossref, https://doi.org/10.1016/j.soildyn.2015.12.009.

Gelman, Andrew, et al. Bayesian data analysis. Chapman and Hall/CRC, 1995.

Geman, Stuart, and Donald Geman. "Stochastic relaxation, Gibbs distributions, and the Bayesian restoration of images." IEEE Transactions on pattern analysis and machine intelligence 6 (1984): 721-741.

Ghani, Sufyan, and Sunita Kumari. "Liquefaction Behavior of Indo-Gangetic Region Using Novel Metaheuristic Optimization Algorithms Coupled with Artificial Neural Network." Natural Hazards, vol. 111, no. 3, Jan. 2022, pp. 2995–3029. Crossref, https://doi.org/10.1007/s11069-021-05165-y.

Ghani, Sufyan, et al. "Soil Categorization and Liquefaction Prediction Using Deep Learning and Ensemble Learning Algorithms." Transportation Infrastructure Geotechnology, vol. 12, no. 1, Nov. 2024. Crossref, https://doi.org/10.1007/s40515-024-00474-8.

Gong, Wenping, et al. "Optimization of Site Investigation Program for Improved Statistical Characterization of Geotechnical Property Based on Random Field Theory." Bulletin of Engineering Geology and the Environment, vol. 76, no. 3, Apr. 2016, pp. 1021–35. Crossref, https://doi.org/10.1007/s10064-016-0869-3.

Gui, Shengxiang, et al. "Probabilistic Slope Stability Analysis with Stochastic Soil Hydraulic Conductivity." Journal of Geotechnical and Geoenvironmental Engineering, vol. 126, no. 1, Jan. 2000, pp. 1–9. Crossref, https://doi.org/10.1061/(asce)1090-0241(2000)126:1(1).

Hamada, M., et al. "Study on Permanent Ground Displacement Induced by Seismic Liquefaction." Computers and Geotechnics, vol. 4, no. 4, Jan. 1987, pp. 197–220. Crossref, https://doi.org/10.1016/0266-352x(87)90001-2.

Hasancebi, Nilsun, and Resat Ulusay. "Empirical Correlations between Shear Wave Velocity and Penetration Resistance for Ground Shaking Assessments." Bulletin of Engineering Geology and the Environment, vol. 66, no. 2, Aug. 2006, pp. 203–13. Crossref, https://doi.org/10.1007/s10064-006-0063-0.

Hastings, W. K. "Monte Carlo Sampling Methods Using Markov Chains and Their Applications." Biometrika, vol. 57, no. 1, Apr. 1970, pp. 97–109. Crossref, https://doi.org/10.1093/biomet/57.1.97.

Hatanaka, Munenori, et al. "Liquefaction Characteristics of A Gravelly Fill Liquefied During the 1995 Hyogo-Ken Nanbu Earthquake." Soils and Foundations, vol. 37, no. 3, Sept. 1997, pp. 107–15. Crossref, https://doi.org/10.3208/sandf.37.3_107.

Hou, Shaokang, et al. "Real-Time Prediction of Rock Mass Classification Based on TBM Operation Big Data and Stacking Technique of Ensemble Learning." Journal of Rock Mechanics and Geotechnical Engineering, vol. 14, no. 1, Feb. 2022, pp. 123–43. Crossref, https://doi.org/10.1016/j.jrmge.2021.05.004.

HOULSBY, N. M. T., and G. T. HOULSBY. "Statistical Fitting of Undrained Strength Data." Géotechnique, vol. 63, no. 14, Nov. 2013, pp. 1253–63. Crossref, https://doi.org/10.1680/geot.13.p.007.

Hsein Juang, C., et al. "Bayesian Updating of Soil Parameters for Braced Excavations Using Field Observations." Journal of Geotechnical and Geoenvironmental Engineering, vol. 139, no. 3, Mar. 2013, pp. 395–406. Crossref, https://doi.org/10.1061/(asce)gt.1943-5606.0000782.

Hu, Jilei, and Huabei Liu. "Bayesian Network Models for Probabilistic Evaluation of Earthquake-Induced Liquefaction Based on CPT and Vs Databases." Engineering Geology, vol. 254, May 2019, pp. 76–88. Crossref, https://doi.org/10.1016/j.enggeo.2019.04.003.

Hu, Ji-Lei, and Hua-Bei Liu. "The Uplift Behaviour of a Subway Station During Different Degrees of Soil Liquefaction." Procedia Engineering, vol. 189, 2017, pp. 18–24. Crossref, https://doi.org/10.1016/j.proeng.2017.05.004.

Hu, Ji-Lei, et al. "Assessment of Seismic Liquefaction Potential Based on Bayesian Network Constructed from Domain Knowledge and History Data." Soil Dynamics and Earthquake Engineering, vol. 89, Oct. 2016, pp. 49–60. Crossref, https://doi.org/10.1016/j.soildyn.2016.07.007.

Hu, Yue, and Yu Wang. "Probabilistic Soil Classification and Stratification in a Vertical Cross-Section from Limited Cone Penetration Tests Using Random Field and Monte Carlo Simulation." Computers and Geotechnics, vol. 124, Aug. 2020, p. 103634. Crossref, https://doi.org/10.1016/j.compgeo.2020.103634.

Hu, Yue, et al. "Bayesian Supervised Learning of Site-Specific Geotechnical Spatial Variability from Sparse Measurements." ASCE-ASME Journal of Risk and Uncertainty in Engineering Systems, Part A: Civil Engineering, vol. 6, no. 2, June 2020. Crossref, https://doi.org/10.1061/ajrua6.0001059.

Huang, Jinsong, et al. "Probabilistic Characterization of Two-Dimensional Soil Profile by Integrating Cone Penetration Test (CPT) with Multi-Channel Analysis of Surface Wave (MASW) Data." Canadian Geotechnical Journal, vol. 55, no. 8, Aug. 2018, pp. 1168–81. Crossref, https://doi.org/10.1139/cgj-2017-0429.

Huang, M., F. Liang, and Z. Li. "Recent advances in the analysis of pile foundation in China." Advances in deep foundations (2007): 127-136.Crossref, https://doi.org/10.1201/9780203938416-10.

Imai, T., and Y. Yoshimura. "Elastic wave velocity and soil properties in soft soil." Tsuchito-kiso 18.1 (1970): 17-22.

Ishihara, Kenji, and Yasuyuki Koga. "Case Studies of Liquefaction in the 1964 Niigata Earthquake." Soils and Foundations, vol. 21, no. 3, Sept. 1981, pp. 35–52. Crossref, https://doi.org/10.3208/sandf1972.21.3_35.

Jas, Kaushik, and G. R. Dodagoudar. "Liquefaction potential assessment of soils using machine learning techniques: a state-of-the-art review from 1994–2021." *International Journal of Geomechanics* 23.7 (2023): 03123002 https://doi.org/10.1061/ijgnai.gmeng-7788.

Jas, Kaushik, et al. "Evaluation and Analysis of Liquefaction Potential of Gravelly Soils Using Explainable Probabilistic Machine Learning Model." Computers and Geotechnics, vol. 167, Mar. 2024, p. 106051. Crossref, https://doi.org/10.1016/j.compgeo.2023.106051.

Jas, Kaushik, et al. "Evaluation and Future Prospects of Data-Driven Intelligence-Based Framework for Predicting Cyclic Behavior of Reconstituted Sand." International Journal for Numerical and Analytical Methods in Geomechanics, vol. 49, no. 5, Jan. 2025, pp. 1597–621. Portico, Crossref, https://doi.org/10.1002/nag.3939.

Jin, Yin-Fu, et al. "Bayesian Model Selection for Sand with Generalization Ability Evaluation." International Journal for Numerical and Analytical Methods in Geomechanics, vol. 43, no. 14, July 2019, pp. 2305–27. Portico, Crossref, https://doi.org/10.1002/nag.2979.

Jin, Yingyan, et al. "A Bayesian Definition of 'Most Probable' Parameters." Geotechnical Research, vol. 5, no. 3, Sept. 2018, pp. 130–42. Crossref, https://doi.org/10.1680/jgere.18.00027.

Johnson, Alicia A., et al. "Bayes' Rule." Bayes Rules!, Jan. 2022, pp. 17–48. Crossref, https://doi.org/10.1201/9780429288340-2.

Jorne, Fernando, et al. "Influence of Superplasticiser, Temperature, Resting Time and Injection Pressure on Hydraulic Lime Grout Injectability. Correlation Analysis between Fresh Grout Parameters and Grout Injectability." Journal of Building Engineering, vol. 4, Dec. 2015, pp. 140–51. Crossref, https://doi.org/10.1016/j.jobe.2015.08.007.

Juang, C. Hsein, et al. "Assessing SPT-Based Probabilistic Models for Liquefaction Potential Evaluation: A 10-Year Update." Georisk: Assessment and Management of Risk for Engineered Systems and Geohazards, vol. 7, no. 3, Sept. 2013, pp. 137–50. Crossref, https://doi.org/10.1080/17499518.2013.778117.

Juang, C. Hsein, et al. "Probabilistic Framework for Liquefaction Potential by Shear Wave Velocity." Journal of Geotechnical and Geoenvironmental Engineering, vol. 127, no. 8, Aug. 2001, pp. 670–78. Crossref, https://doi.org/10.1061/(asce)1090-0241(2001)127:8(670).

Kalinli, Adem, et al. "New Approaches to Determine the Ultimate Bearing Capacity of Shallow Foundations Based on Artificial Neural Networks and Ant Colony Optimization." Engineering Geology, vol. 117, no. 1–2, Jan. 2011, pp. 29–38. Crossref, https://doi.org/10.1016/j.enggeo.2010.10.002.

Kardani, Navid, et al. "Estimation of Bearing Capacity of Piles in Cohesionless Soil Using Optimised Machine Learning Approaches." Geotechnical and Geological Engineering, vol. 38, no. 2, Nov. 2019, pp. 2271–91. Crossref, https://doi.org/10.1007/s10706-019-01085-8.

Karol, Reuben H. Chemical Grouting And Soil Stabilisation, Revised And Expanded. CRC Press, 2003. Crossref, https://doi.org/10.1201/9780203911815.

Karthikeyan, J., and Pijush Samui. "Application of Statistical Learning Algorithms for Prediction of Liquefaction Susceptibility of Soil Based on Shear Wave Velocity." Geomatics, Natural Hazards and Risk, vol. 5, no. 1, Jan. 2013, pp. 7–25. Crossref, https://doi.org/10.1080/19475705.2012.757252.

Kayen, R., et al. "Shear-Wave Velocity–Based Probabilistic and Deterministic Assessment of Seismic Soil Liquefaction Potential." Journal of Geotechnical and Geoenvironmental Engineering, vol. 139, no. 3, Mar. 2013, pp. 407–19. Crossref, https://doi.org/10.1061/(asce)gt.1943-5606.0000743.

Kayen, R., et al. "Shear-Wave Velocity–Based Probabilistic and Deterministic Assessment of Seismic Soil Liquefaction Potential." Journal of Geotechnical and Geoenvironmental Engineering, vol. 139, no. 3, Mar. 2013, pp. 407–19. Crossref, https://doi.org/10.1061/(asce)gt.1943-5606.0000743.

Kayen, Robert E., and James K. Mitchell. "Assessment of Liquefaction Potential during Earthquakes by Arias Intensity." Journal of Geotechnical and Geoenvironmental Engineering, vol. 123, no. 12, Dec. 1997, pp. 1162–74. Crossref, https://doi.org/10.1061/(asce)1090-0241(1997)123:12(1162).

Khatti, Jitendra, et al. "Cone Penetration Test-Based Assessment of Liquefaction Potential Using Machine and Hybrid Learning Approaches." Multiscale and Multidisciplinary Modeling, Experiments and Design, vol. 7, no. 4, Apr. 2024, pp. 3841–64. Crossref, https://doi.org/10.1007/s41939-024-00447-x.

Kiranyaz, Serkan, et al. "Evolutionary Artificial Neural Networks by Multi-Dimensional Particle Swarm Optimization." Neural Networks, vol. 22, no. 10, Dec. 2009, pp. 1448–62. Crossref, https://doi.org/10.1016/j.neunet.2009.05.013.

Kirar, Bablu, et al. "Correlation Between Shear Wave Velocity (Vs) and SPT Resistance (N) for Roorkee Region." International Journal of Geosynthetics and Ground Engineering, vol. 2, no. 1, Mar. 2016. Crossref, https://doi.org/10.1007/s40891-016-0047-5.

Kramer, Steven L., and Jonathan P. Stewart. "Introduction to Geotechnical Earthquake Engineering." Geotechnical Earthquake Engineering, Oct. 2024, pp. 1–26. Crossref, https://doi.org/10.1201/9781003512011-1.

Kulhawy, Fred H. "On the evaluation of static soil properties." Stability and performance of slopes and embankments II. ASCE, 1993.

Kumar, Deepak, et al. "A Novel Methodology to Classify Soil Liquefaction Using Deep Learning." Geotechnical and Geological Engineering, vol. 39, no. 2, Sept. 2020, pp. 1049–58. Crossref, https://doi.org/10.1007/s10706-020-01544-7.

Kurnaz, Talas Fikret, et al. "A Hyper Parameterized Artificial Neural Network Approach for Prediction of the Factor of Safety against Liquefaction." Engineering Geology, vol. 319, June 2023, p. 107109. Crossref, https://doi.org/10.1016/j.enggeo.2023.107109.

Lacasse, Suzanne, et al. "Probabilistic Characterization of Soil Properties—Recognition of Wilson Tang's Contribution to Geotechnical Practice." Geotechnical Safety and Reliability, June 2017, pp. 2–26. Crossref, https://doi.org/10.1061/9780784480731.002.

Langseth, Helge, et al. "Inference in Hybrid Bayesian Networks." Reliability Engineering & System Safety, vol. 94, no. 10, Oct. 2009, pp. 1499–509. Crossref, https://doi.org/10.1016/j.ress.2009.02.027.

Larrañaga, P., and S. Moral. "Probabilistic Graphical Models in Artificial Intelligence." Applied Soft Computing, vol. 11, no. 2, Mar. 2011, pp. 1511–28. Crossref, https://doi.org/10.1016/j.asoc.2008.01.003.

Li, Cong, et al. "Bayesian Approach for Sequential Probabilistic Back Analysis of Uncertain Geomechanical Parameters and Reliability Updating of Tunneling-Induced Ground Settlements." Advances in Civil Engineering, edited by Paolo Castaldo, vol. 2020, no. 1, Jan. 2020. Portico, Crossref, https://doi.org/10.1155/2020/8528304.

Li, Enming, et al. "Prediction of Blasting Mean Fragment Size Using Support Vector Regression Combined with Five Optimization Algorithms." Journal of Rock Mechanics and Geotechnical Engineering, vol. 13, no. 6, Dec. 2021, pp. 1380–97. Crossref, https://doi.org/10.1016/j.jrmge.2021.07.013.

Li, J., et al. "Bayesian Prediction of Peak Resistance of a Spudcan Penetrating Sand-over-Clay." Géotechnique, vol. 68, no. 10, Oct. 2018, pp. 905–17. Crossref, https://doi.org/10.1680/jgeot.17.p.154.

Liao, Kuo-Wei, et al. "An Artificial Neural Network for Groutability Prediction of Permeation Grouting with Microfine Cement Grouts." Computers and Geotechnics, vol. 38, no. 8, Dec. 2011, pp. 978–86. Crossref, https://doi.org/10.1016/j.compgeo.2011.07.008.

Lin, Shan, et al. "Comparative Performance of Eight Ensemble Learning Approaches for the Development of Models of Slope Stability Prediction." Acta Geotechnica, vol. 17, no. 4, Jan. 2022, pp. 1477–502. Crossref, https://doi.org/10.1007/s11440-021-01440-1.

Liu, Likai, and Zechun Hu. "Data-Driven Regulation Reserve Capacity Determination Based on Bayes Theorem." IEEE Transactions on Power Systems, vol. 35, no. 2, Mar. 2020, pp. 1646–49. Crossref, https://doi.org/10.1109/tpwrs.2020.2965763.

Liu, Songyu, et al. "Multivariate Correlation among Resilient Modulus and Cone Penetration Test Parameters of Cohesive Subgrade Soils." Engineering Geology, vol. 209, July 2016, pp. 128–42. Crossref, https://doi.org/10.1016/j.enggeo.2016.05.018.

Lunn, David, et al. The BUGS Book. Chapman and Hall/CRC, 2012. Crossref, https://doi.org/10.1201/b13613.

mai, Tsuneo, and Keiji Tonoughi. "Correlation of N Value with S-Wave Velocity and Shear Modulus." Penetration Testing, Feb. 2021, pp. 67–72. Crossref, https://doi.org/10.1201/9780203743959-11.

Mangalathu, Sujith, et al. "Rapid Seismic Damage Evaluation of Bridge Portfolios Using Machine Learning Techniques." Engineering Structures, vol. 201, Dec. 2019, p. 109785. Crossref, https://doi.org/10.1016/j.engstruct.2019.109785.

Markou, I. N., et al. "Injectability of Microfine Cement Grouts into Limestone Sands with Different Gradations: Experimental Investigation and Prediction." Geotechnical and Geological Engineering, Sept. 2017. Crossref, https://doi.org/10.1007/s10706-017-0368-8.

Metropolis, Nicholas, et al. "Equation of State Calculations by Fast Computing Machines." The Journal of Chemical Physics, vol. 21, no. 6, June 1953, pp. 1087–92. Crossref, https://doi.org/10.1063/1.1699114.

Miro, S., et al. "A Probabilistic Analysis of Subsoil Parameters Uncertainty Impacts on Tunnel-Induced Ground Movements with a Back-Analysis Study." Computers and Geotechnics, vol. 68, July 2015, pp. 38–53. Crossref, https://doi.org/10.1016/j.compgeo.2015.03.012.

Mohammed, Mariamme, et al. "Shallow Foundation Settlement Quantification: Application of Hybridized Adaptive Neuro-Fuzzy Inference System Model." Advances in Civil Engineering, edited by Castorina S. Vieira, vol. 2020, no. 1, Jan. 2020. Portico, Crossref, https://doi.org/10.1155/2020/7381617.

Molnar, S., et al. "A Review of the Microtremor Horizontal-to-Vertical Spectral Ratio (MHVSR) Method." Journal of Seismology, vol. 26, no. 4, Mar. 2022, pp. 653–85. Crossref, https://doi.org/10.1007/s10950-021-10062-9.

Murphy, Kevin P. Machine learning: a probabilistic perspective. MIT press, 2012.

Ohta, Yutaka, et al. "Shear Wave Velocity Measurement during a Standard Penetration Test." Earthquake Engineering & Structural Dynamics, vol. 6, no. 1, Jan. 1978, pp. 43–50. Portico, Crossref, https://doi.org/10.1002/eqe.4290060106.

Opitz, D., and R. Maclin. "Popular Ensemble Methods: An Empirical Study." Journal of Artificial Intelligence Research, vol. 11, Aug. 1999, pp. 169–98. Crossref, https://doi.org/10.1613/jair.614.

Ozgurel, H. Gurkan, and Cumaraswamy Vipulanandan. "Effect of Grain Size and Distribution on Permeability and Mechanical Behavior of Acrylamide Grouted Sand." Journal of Geotechnical and Geoenvironmental Engineering, vol. 131, no. 12, Dec. 2005, pp. 1457–65. Crossref, https://doi.org/10.1061/(asce)1090-0241(2005)131:12(1457).

Padmini, D., et al. "Ultimate Bearing Capacity Prediction of Shallow Foundations on Cohesionless Soils Using Neurofuzzy Models." Computers and Geotechnics, vol. 35, no. 1, Jan. 2008, pp. 33–46. Crossref, https://doi.org/10.1016/j.compgeo.2007.03.001.

Papaioannou, Iason, and Daniel Straub. "Reliability Updating in Geotechnical Engineering Including Spatial Variability of Soil." Computers and Geotechnics, vol. 42, May 2012, pp. 44–51. Crossref, https://doi.org/10.1016/j.compgeo.2011.12.004.

Peck, R. B. "Advantages and Limitations of the Observational Method in Applied Soil Mechanics." Géotechnique, vol. 19, no. 2, June 1969, pp. 171–87. Crossref, https://doi.org/10.1680/geot.1969.19.2.171.

Peck, Ralph B. "Liquefaction Potential: Science Versus Practice." Journal of the Geotechnical Engineering Division, vol. 105, no. 3, Mar. 1979, pp. 393–98. Crossref, https://doi.org/10.1061/ajgeb6.0000776.

Perret, S., et al. "Repair of 130-Year Old Masonry Bridge Using High-Performance Cement Grout." Journal of Bridge Engineering, vol. 7, no. 1, Jan. 2002, pp. 31–38. Crossref, https://doi.org/10.1061/(asce)1084-0702(2002)7:1(31).

Perret, Stephanie, Danielle Palardy, and Gerard Ballivy. "Rheological Behaviour and Setting Time of Microfine Cement-Based Grouts." ACI Materials Journal, vol. 97, no. 4, 2000. Crossref, https://doi.org/10.14359/7413.

Phoon, Kok-Kwang, and Fred H. Kulhawy. "Characterisation of Geotechnical Variability." Canadian Geotechnical Journal, vol. 36, no. 4, Nov. 1999, pp. 612–24. Crossref, https://doi.org/10.1139/t99-038.

Phoon, Kok-Kwang, and Wengang Zhang. "Future of Machine Learning in Geotechnics." Georisk: Assessment and Management of Risk for Engineered Systems and Geohazards, vol. 17, no. 1, June 2022, pp. 7–22. Crossref, https://doi.org/10.1080/17499518.2022.2087884.

Phoon, Kok-Kwang, et al. "Managing Risk in Geotechnical Engineering – From Data to Digitalisation." Proceedings of the 7th International Symposium on Geotechnical Safety and Risk (ISGSR 2019), 2019, pp. 13–34. Crossref, https://doi.org/10.3850/978-981-11-2725-0-sl-cd.

Phoon, Kok-Kwang. "Role of Reliability Calculations in Geotechnical Design." Georisk: Assessment and Management of Risk for Engineered Systems and Geohazards, vol. 11, no. 1, Dec. 2016, pp. 4–21. https://doi.org/10.1080/17499518.2016.1265653.

Popescu, Radu, and Jean H. Prevost. "Reliability Assessment of Centrifuge Soil Test Results." Soil Dynamics and Earthquake Engineering, vol. 14, no. 2, Jan. 1995, pp. 93–101. Crossref, https://doi.org/10.1016/0267-7261(94)00037-h.

Qi, Xiao-Hui, and Wan-Huan Zhou. "An Efficient Probabilistic Back-Analysis Method for Braced Excavations Using Wall Deflection Data at Multiple Points." Computers and Geotechnics, vol. 85, May 2017, pp. 186–98. Crossref, https://doi.org/10.1016/j.compgeo.2016.12.032.

Rahimi, Salman, et al. "Influence of Soil Aging on SPT-Vs Correlation and Seismic Site Classification." Engineering Geology, vol. 272, July 2020, p. 105653. Crossref, https://doi.org/10.1016/j.enggeo.2020.105653.

Sahin, Emrehan Kutlug, and Selcuk Demir. "Greedy-AutoML: A Novel Greedy-Based Stacking Ensemble Learning Framework for Assessing Soil Liquefaction Potential." Engineering Applications of Artificial Intelligence, vol. 119, Mar. 2023, p. 105732. Crossref, https://doi.org/10.1016/j.engappai.2022.105732.

Santagata, M. C., and E. Santagata. "Experimental Investigation of Factors Affecting the Injectability of Microcement Grouts." Grouting and Ground Treatment, Jan. 2003, pp. 1221–34. Crossref, https://doi.org/10.1061/40663(2003)70.

Sassa, Shinji, and Tomohiro Takagawa. "Liquefied Gravity Flow-Induced Tsunami: First Evidence and Comparison from the 2018 Indonesia Sulawesi Earthquake and Tsunami Disasters." Landslides, vol. 16, no. 1, Nov. 2018, pp. 195–200. Crossref, https://doi.org/10.1007/s10346-018-1114-x.

Seed, H. Bolton, and Izzat M. Idriss. "Simplified Procedure for Evaluating Soil Liquefaction Potential." Journal of the Soil Mechanics and Foundations Division, vol. 97, no. 9, Sept. 1971, pp. 1249–73. Crossref, https://doi.org/10.1061/jsfeaq.0001662.

Seed, H. Bolton, et al. "Closure to 'Evaluation of Liquefaction Potential Using Field Performance Data by H. Bolton Seed, I. M. Idriss, and Ignacio Arango, (March, 1983)." Journal of Geotechnical Engineering, vol. 111, no. 11, Nov. 1985, pp. 1346–1346. Crossref, https://doi.org/10.1061/(asce)0733-9410(1985)111:11(1346).

Seed, H. Bolton, et al. "Evaluation of Liquefaction Potential Using Field Performance Data." Journal of Geotechnical Engineering, vol. 109, no. 3, Mar. 1983, pp. 458–82. Crossref, https://doi.org/10.1061/(asce)0733-9410(1983)109:3(458).

Shen, Mengfen, et al. "Predicting Liquefaction Probability Based on Shear Wave Velocity: An Update." Bulletin of Engineering Geology and the Environment, vol. 75, no. 3, Apr. 2016, pp. 1199–214. Crossref, https://doi.org/10.1007/s10064-016-0880-8.

Shenoy, Prakash P., and James C. West. "Inference in Hybrid Bayesian Networks Using Mixtures of Polynomials." International Journal of Approximate Reasoning, vol. 52, no. 5, July 2011, pp. 641–57. Crossref, https://doi.org/10.1016/j.ijar.2010.09.003.

Shi, Chao, and Yu Wang. "Development of Subsurface Geological Cross-Section from Limited Site-Specific Boreholes and Prior Geological Knowledge Using Iterative Convolution XGBoost." Journal of Geotechnical and Geoenvironmental Engineering, vol. 147, no. 9, Sept. 2021. Crossref, https://doi.org/10.1061/(asce)gt.1943-5606.0002583.

Shi, Chao, and Yu Wang. "Nonparametric and Data-Driven Interpolation of Subsurface Soil Stratigraphy from Limited Data Using Multiple Point Statistics." Canadian Geotechnical Journal, vol. 58, no. 2, Feb. 2021, pp. 261–80. Crossref, https://doi.org/10.1139/cgj-2019-0843.

Shibata, T. "The relationship between the N-value and S-wave velocity in the soil layer." Disaster Prevention Research Laboratory, Kyoto University, Kyoto, Japan (1970).

Spross, Johan, and Stefan Larsson. "On the Observational Method for Groundwater Control in the Northern Link Tunnel Project, Stockholm, Sweden." Bulletin of Engineering Geology and the Environment, vol. 73, no. 2, Jan. 2014, pp. 401–08. Crossref, https://doi.org/10.1007/s10064-013-0501-8.

Spross, Johan, and Stefan Larsson. "Probabilistic Observational Method for Design of Surcharges on Vertical Drains." Géotechnique, Jan. 2021, pp. 1–13. Crossref, https://doi.org/10.1680/jgeot.19.p.053.

Sun, Junlong, et al. "Based on Multi-Algorithm Hybrid Method to Predict the Slope Safety Factor-- Stacking Ensemble Learning with Bayesian Optimization." Journal of Computational Science, vol. 59, Mar. 2022, p. 101587. Crossref, https://doi.org/10.1016/j.jocs.2022.101587.

Sykora, David Wayne. Correlations of in situ measurements in sands of shear wave velocity, soil characteristics, and site conditions. Diss. University of Texas at Austin, 1983.

Tang, Chong, and Kok-Kwang Phoon. Model Uncertainties in Foundation Design. Mar. 2021. Crossref, https://doi.org/10.1201/9780429024993.

Tang, W. H. "Principles of Probabilistic Characterisation of Soil Properties." Geotechnical Safety and Reliability, June 2017, pp. 39–39. Crossref, https://doi.org/10.1061/9780784480731.004.

Tao, Yuanqin, et al. "Hierarchical Bayesian Model for Predicting Small-Strain Stiffness of Sand." Canadian Geotechnical Journal, vol. 61, no. 4, Apr. 2024, pp. 668–83. Crossref, https://doi.org/10.1139/cgj-2022-0598.

Tekin, E., and S. O. Akbas. "Artificial Neural Networks Approach for Estimating the Groutability of Granular Soils with Cement-Based Grouts." Bulletin of Engineering Geology and the Environment, vol. 70, no. 1, May 2010, pp. 153–61. Crossref, https://doi.org/10.1007/s10064-010-0295-x.

Thokchom, Sarda, et al. "Empirical Correlation of SPT Blow Counts versus Shear Wave Velocity for Different Types of Soils in Dholera, Western India." Natural Hazards, vol. 86, no. 3, Jan. 2017, pp. 1291–306. Crossref, https://doi.org/10.1007/s11069-017-2744-3.

Tiwari, Satyam, Sarat K. Das, and Madhumita Mohanty. "Improved prediction of soil liquefaction susceptibility using ensemble learning algorithms." *Geomechanics & engineering* 37.5 (2024): 475-498 https://doi.org/10.12989/gae.2024.37.5.475.

Tung, Albert T. Y., et al. "Assessment of Liquefaction Potential Using Neural Networks." Soil Dynamics and Earthquake Engineering, vol. 12, no. 6, Jan. 1993, pp. 325–35. Crossref, https://doi.org/10.1016/0267-7261(93)90035-p.

Uma Maheswari, R., et al. "Use of Surface Waves in Statistical Correlations of Shear Wave Velocity and Penetration Resistance of Chennai Soils." Geotechnical and Geological Engineering, vol. 28, no. 2, Dec. 2009, pp. 119–37. Crossref, https://doi.org/10.1007/s10706-009-9285-9.

Uusitalo, Laura. "Advantages and Challenges of Bayesian Networks in Environmental Modelling." Ecological Modelling, vol. 203, no. 3–4, May 2007, pp. 312–18. Crossref, https://doi.org/10.1016/j.ecolmodel.2006.11.033.

Uzielli, Marco, et al. "Bayesian Prediction of Punch-Through Probability for Spudcans in Stiff-over-Soft Clay." Geotechnical Safety and Reliability, June 2017, pp. 247–65. Crossref, https://doi.org/10.1061/9780784480731.020.

van de Schoot, Rens, et al. "Bayesian Statistics and Modelling." Nature Reviews Methods Primers, vol. 1, no. 1, Jan. 2021. Crossref, https://doi.org/10.1038/s43586-020-00001-2.

van der Krogt, Mark G., et al. "Improving Dike Reliability Estimates by Incorporating Construction Survival." Engineering Geology, vol. 280, Jan. 2021, p. 105937. Crossref, https://doi.org/10.1016/j.enggeo.2020.105937.

W. Krämer, "Andrew Gelman and Jennifer Hill: Data analysis using regression and multilevel/hierarchical models," Statistical Papers, vol. 52, no. 3, pp. 741–742, Feb. 2010, doi: 10.1007/s00362-010-0311-z.

Wang, Lei, et al. "Bayesian Updating of KJHH Model for Prediction of Maximum Ground Settlement in Braced Excavations Using Centrifuge Data." Computers and Geotechnics, vol. 44, June 2012, pp. 1–8. Crossref, https://doi.org/10.1016/j.compgeo.2012.03.003.

Wang, Lin, et al. "Determination of Site-Specific Soil-Water Characteristic Curve from a Limited Number of Test Data – A Bayesian Perspective." Geoscience Frontiers, vol. 9, no. 6, Nov. 2018, pp. 1665–77. Crossref, https://doi.org/10.1016/j.gsf.2017.10.014.

Wang, Xiangrong, et al. "A Hidden Markov Random Field Model-Based Approach for Probabilistic Site Characterization Using Multiple Cone Penetration Test Data." Structural Safety, vol. 70, Jan. 2018, pp. 128–38. Crossref, https://doi.org/10.1016/j.strusafe.2017.10.011.

Wang, Y., and T. Zhao. "Bayesian Assessment of Site-specific Performance of Geotechnical Design Charts with Unknown Model Uncertainty." International Journal for Numerical and Analytical Methods in Geomechanics, vol. 41, no. 5, Nov. 2016, pp. 781–800. Portico, Crossref, https://doi.org/10.1002/nag.2658.

Wang, Yu, and Adeyemi Emman Aladejare. "Bayesian Characterisation of Correlation between Uniaxial Compressive Strength and Young's Modulus of Rock." International Journal of Rock Mechanics and Mining Sciences, vol. 85, May 2016, pp. 10–19. Crossref, https://doi.org/10.1016/j.ijrmms.2016.02.010.

Wang, Yu, and Oluwatosin Victor Akeju. "Quantifying the Cross-Correlation between Effective Cohesion and Friction Angle of Soil from Limited Site-Specific Data." Soils and Foundations, vol. 56, no. 6, Dec. 2016, pp. 1055–70. Crossref, https://doi.org/10.1016/j.sandf.2016.11.009.

Wang, Yu, and Zijun Cao. "Probabilistic Characterization of Young's Modulus of Soil Using Equivalent Samples." Engineering Geology, vol. 159, June 2013, pp. 106–18. Crossref, https://doi.org/10.1016/j.enggeo.2013.03.017.

Wang, Yu, et al. "Bayesian Perspective on Geotechnical Variability and Site Characterization." Engineering Geology, vol. 203, Mar. 2016, pp. 117–25. Crossref, https://doi.org/10.1016/j.enggeo.2015.08.017.

Wang, Yu, et al. "Non-Parametric Modelling and Simulation of Spatiotemporally Varying Geo-Data." Georisk: Assessment and Management of Risk for Engineered Systems and Geohazards, vol. 16, no. 1, Sept. 2021, pp. 77–97. Crossref, https://doi.org/10.1080/17499518.2021.1971258.

Wang, Yu, et al. "Site-Specific Probability Distribution of Geotechnical Properties." Computers and Geotechnics, vol. 70, Oct. 2015, pp. 159–68. Crossref, https://doi.org/10.1016/j.compgeo.2015.08.002.

Wang, Z. Z., and C. Chen. "Fuzzy Comprehensive Bayesian Network-Based Safety Risk Assessment for Metro Construction Projects." Tunnelling and Underground Space Technology, vol. 70, Nov. 2017, pp. 330–42. Crossref, https://doi.org/10.1016/j.tust.2017.09.012.

Wu, Stephen, et al. "Quasi-Site-Specific Soil Property Prediction Using a Cluster-Based Hierarchical Bayesian Model." Structural Safety, vol. 99, Nov. 2022, p. 102253. Crossref, https://doi.org/10.1016/j.strusafe.2022.102253.

Wu, Tien H. "2008 Peck Lecture: The Observational Method: Case History and Models." Journal of Geotechnical and Geoenvironmental Engineering, vol. 137, no. 10, Oct. 2011, pp. 862–73. Crossref, https://doi.org/10.1061/(asce)gt.1943-5606.0000509.

Xiao, Shihao, et al. "Establishing Region-Specific N – V Relationships through Hierarchical Bayesian Modeling." Engineering Geology, vol. 287, June 2021, p. 106105. Crossref, https://doi.org/10.1016/j.enggeo.2021.106105.

Yang, Hao-Qing, et al. "Efficient Method for Probabilistic Estimation of Spatially Varied Hydraulic Properties in a Soil Slope Based on Field Responses: A Bayesian Approach." Computers and Geotechnics, vol. 102, Oct. 2018, pp. 262–72. Crossref, https://doi.org/10.1016/j.compgeo.2017.11.012.

Yang, Lingqiang, et al. "Predicting the Young's Modulus of Granites Using the Bayesian Model Selection Approach." Bulletin of Engineering Geology and the Environment, vol. 78, no. 5, June 2018, pp. 3413–23. Crossref, https://doi.org/10.1007/s10064-018-1326-2.

Yu, Wei Wei, and Xuan Guo. "Study of Ground Lateral Flow on Pile Foundation Due to Soil Liquefaction." Advanced Materials Research, vol. 415–417, Dec. 2011, pp. 1478–81. Crossref, https://doi.org/10.4028/www.scientific.net/amr.415-417.1478.

Yuan, Haiming, et al. "Liquefaction-Induced Ground Failure: A Study of the Chi-Chi Earthquake Cases." Engineering Geology, vol. 71, no. 1–2, Jan. 2004, pp. 141–55. Crossref, https://doi.org/10.1016/s0013-7952(03)00130-3.

Yu-Kun Bao, et al. "Forecasting Stock Composite Index by Fuzzy Support Vector Machines Regression." 2005 International Conference on Machine Learning and Cybernetics, 2005, pp. 3535-3540 Vol. 6. Crossref, https://doi.org/10.1109/icmlc.2005.1527554.

Zebovitz, S., et al. "Injection of Fine Sands with Very Fine Cement Grout." Journal of Geotechnical Engineering, vol. 115, no. 12, Dec. 1989, pp. 1717–33. Crossref, https://doi.org/10.1061/(asce)0733-9410(1989)115:12(1717).

Zhang, Dong-Liang, et al. "Cyclic Behaviour and Ultimate Bearing Capacity of Circular Concrete-Filled Double Skin Steel Tube Members Subjected to Combined Compression and Torsion." Thin-Walled Structures, vol. 186, May 2023, p. 110707. Crossref, https://doi.org/10.1016/j.tws.2023.110707.

Zhang, Jie, et al. "Bayesian Network Based Machine Learning for Design of Pile Foundations." Automation in Construction, vol. 118, Oct. 2020, p. 103295. Crossref, https://doi.org/10.1016/j.autcon.2020.103295.

Zhang, Junfei, et al. "Strength of Ensemble Learning in Multiclass Classification of Rockburst Intensity." International Journal for Numerical and Analytical Methods in Geomechanics, vol. 44, no. 13, June 2020, pp. 1833–53. Portico, Crossref, https://doi.org/10.1002/nag.3111.

Zhang, L. L., et al. "Uncertainties of Field Pullout Resistance of Soil Nails." Journal of Geotechnical and Geoenvironmental Engineering, vol. 135, no. 7, July 2009, pp. 966–72. Crossref, https://doi.org/10.1061/(asce)gt.1943-5606.0000014.

Zhang, Limao, et al. "Bayesian-Network-Based Safety Risk Analysis in Construction Projects." Reliability Engineering & System Safety, vol. 131, Nov. 2014, pp. 29–39. Crossref, https://doi.org/10.1016/j.ress.2014.06.006.

Zhang, Limin, Y. R. Zheng, and J. L. Wang. "Errors in calculating hydrodynamic pressures for stability analysis of soil slopes subject to rainfall." Proc. Ninth International Conference on Applications of Statistics and Probability in Civil Engineering, ICASP9, San Francisco, USA. 2003.

Zhang, Ruiliang, and Xinhua Xue. "Determining Ultimate Bearing Capacity of Shallow Foundations by Using Multi Expression Programming (MEP)." Engineering Applications of Artificial Intelligence, vol. 115, Oct. 2022, p. 105255. Crossref, https://doi.org/10.1016/j.engappai.2022.105255.

Zhang, Runhong, et al. "Analysis of Ground Surface Settlement in Anisotropic Clays Using Extreme Gradient Boosting and Random Forest Regression Models." Journal of Rock Mechanics and Geotechnical Engineering, vol. 13, no. 6, Dec. 2021, pp. 1478–84. Crossref, https://doi.org/10.1016/j.jrmge.2021.08.001.

Zhang, Wengang, et al. "Application of Deep Learning Algorithms in Geotechnical Engineering: A Short Critical Review." Artificial Intelligence Review, vol. 54, no. 8, Feb. 2021, pp. 5633–73. Crossref, https://doi.org/10.1007/s10462-021-09967-1.

Zhao, Tengyuan, et al. "Derivation of Site-Specific Soil-Water Characteristic Curve (SWCC) from Extremely Sparse Experimental Data by Hierarchical Bayesian Method with Consideration of Geotechnical Sites Similarity." Engineering Geology, vol. 342, Nov. 2024, p. 107752. Crossref, https://doi.org/10.1016/j.enggeo.2024.107752.

Zheng, Dong, et al. "Embankment Prediction Using Testing Data and Monitored Behaviour: A Bayesian Updating Approach." Computers and Geotechnics, vol. 93, Jan. 2018, pp. 150–62. Crossref, https://doi.org/10.1016/j.compgeo.2017.05.003.

Zhou, Wan-Huan, et al. "Estimation of Maximum Pullout Shear Stress of Grouted Soil Nails Using Bayesian Probabilistic Approach." International Journal of Geomechanics, vol. 13, no. 5, Oct. 2013, pp. 659–64. Crossref, https://doi.org/10.1061/(asce)gm.1943-5622.0000259.

Zhou, Ying, et al. "Using Bayesian Network for Safety Risk Analysis of Diaphragm Wall Deflection Based on Field Data." Reliability Engineering & System Safety, vol. 180, Dec. 2018, pp. 152–67. Crossref, https://doi.org/10.1016/j.ress.2018.07.014.

CHAPTER

3

Bayesian Multilevel Modelling for V_s and N relationship in Sandy Soil

3.1 Introduction

Shear wave velocity (*Vs*) is crucial in soil engineering, indicating soil behaviour under dynamic forces like earthquakes. It helps determine stiffness, moisture content, and settlement potential and is fundamental for earthquake hazard analysis. Establishing a correlation between Shear Wave Velocity (*Vs*) and Standard Penetration Test (SPT) blow count (*N*) in the sandy soil is crucial for geotechnical engineering. This link bridges the gap between a readily obtainable in-situ test (SPT) and a critical dynamic soil property (*Vs*). Vs directly influences seismic response and liquefaction potential – vital considerations for safe foundation design in earthquake-prone regions. By correlating *N* with *Vs*, engineers can estimate the latter without resorting to expensive or time-consuming geophysical testing in sandy soils. This not only streamlines site investigations but also facilitates the development of regional Vs maps for improved seismic hazard assessments. (Adamo et al. 2010; Tong et al. 2018; Huang and Zhang 2021; Wair et al. 2012; Zhang et al. 2021). Direct geophysical measurements of Vs are accurate but costly. To address this, engineers often use the standard penetration test (SPT), which provides a blow count (*N*) correlated with *Vs* (Oliveira et al. 2020). Fooladi et al. (2024) used standard penetration test and shear wave velocity data for reliability analysis of liquefaction potential. Johari (2019) used an analytical approach, probabilistic modelling of liquefaction based on shear wave velocity. Obtaining a formulated correlation between *N* and V_s started in Japan during the 1960s, such as those by Kanai et al. (1966), laid the foundation for understanding the correlation between (V_s) and (*N*). This work was expanded globally through seismic microzonation studies on sandy soils, leading to region-specific relationships (Fabbrocino et al. 2015).

The objectives of this study are threefold. First, to test the hypothesis that the slope remains relatively constant for sandy soils, while the intercept varies depending on the region where the sandy soil is located, using a Bayesian multilevel modelling approach. Second, to investigate the use of Bayesian multilevel modelling in characterising the variability in sandy soils. Lastly, to evaluate the potential applicability of the developed model for predicting parameters in new, previously unexamined regions.

3.2. Methodology and data

The V_s-N relationship in sandy soils proposed by various researchers is presented in Table 3.1.

Table 3.1. Previous region-specific studies of the V_s-N relationship in sandy soils

Reference	V_s-N relationship in sandy soil
Daag et al. (2022)	$45.07N^{0.5535}$
Kirar et al. (2016)	$100.3N^{0.338}$
Fatehnia et al. (2018)	$77.1N^{0.355}$
Esfehanizadeh et al.(2015)	$107.2N^{0.34}$
Maheshwari et al.(2013)	$95N^{0.30}$
Maheswari et al.(2010)	$100.53N^{0.265}$
Dikmen et al.(2009)	$73N^{0.33}$
Hanumantharao and Ramana (2008)	$79N^{0.434}$
Hasancebi and Ulusay 2007	$90.8N^{0.319}$
Jafari et al. (2002)	$80N^{0.33}$
Raptakis et al.(1995)	$123.4N^{0.29}$
Pitilakis et al.(1992)	$162N^{0.17}$
Lee, S.H._H. (1992)	$157N^{0.49}$
Okamoto et al.(1989)	$125N^{0.3}$
Sykora et al.(1983)	$100N^{0.29}$
Seed et al.(1983)	$56N^{0.5}$
Japan Road Association (1980)	$80N^{0.333}$
Ohta (1978)	$88N^{0.34}$
Ohsaki et al.(1973)	$59N^{0.47}$
Shibata (1970)	$32N^{0.5}$

These studies typically use a power-law relationship in sandy soil as,

$$V_s = aN^b \tag{3.1}$$

With a and b as fitting parameters derived from regional data. Despite extensive research on N-V_s relationships, significant gaps remain in understanding the variability of the relationship between (V_s) and (N) in sandy soil. The variability in the Vs-N relationships in sandy soil across different studies can be attributed to several factors. Primarily, regional differences in soil composition and geotechnical properties, such as grain size distribution, density, and mineralogy, influence the shear wave velocity (Vs) and standard penetration test (N) values. Additionally, methodological variations in testing procedures, sample handling, and equipment calibration can lead to discrepancies. Environmental conditions, such as moisture content and confining pressure during testing, also play a critical role. Furthermore, empirical models are often site-specific, tailored to local conditions, and may not be universally applicable, resulting in differing *Vs-N* relationships. Current models often overlook regional differences. Generally, data are collected from a region and using regression, the correlation is obtained, necessitating the application of multilevel modelling to quantify the impact of regional variability on the N-V_s relationship. Multilevel modelling (MLM) has not been fully explored to address the calibration of the *Vs-N* relationship for sandy soil. Additionally, existing studies are limited by small datasets, underscoring the need for larger, more diverse datasets to enhance model robustness. Also, a thorough comparative analysis of frequentist and Bayesian MLM frameworks is required to identify the most effective approach for handling variability and uncertainty. Multilevel modelling, particularly through the hierarchical Bayesian model (HBM), offers a promising solution for this challenge. Unlike traditional regression methods that assume data points are independent, multilevel modelling acknowledges that data can be grouped into clusters (e.g., geographical regions). For instance, N-V_s data from different regions can be treated as clusters, each with its own mean and variability. At the first level, data within each cluster (region) are analysed to understand local variability. At the second level, the model accounts for variability between clusters, allowing for the sharing of information across regions. This hierarchical structure enables the model to make more accurate predictions even in regions with limited data by borrowing strength from other regions with abundant data (e.g., Gelman and Shalizi 2013; Zhang et al. 2014). Xiao et at (2021) developed a generalised method using the Hierarchical Bayesian Model (HBM) to establish N-V_s relationships in regions lacking sufficient region-specific data. It outlines a comprehensive procedure from a method suggestion to verification and comparison with alternative models, aiming to provide a practical solution for data-scarce regions. In contrast, this research is focused on the empirical examination and establishment of the N-V_s relationship specifically for sandy soils from various geographical regions. It employs a Bayesian multilevel approach to investigate and characterise the variability of N-V_s data in sandy soils and assess the types of predictions that can be made based on the established relationships, and compare results from different pooling of data (complete pooling, no pooling, partial pooling).

The Bayesian approach further enhances this modelling by treating parameters as random variables, allowing for the incorporation of prior knowledge and the quantification of uncertainty in a probabilistic framework. This results in probability distributions for parameters, providing a more nuanced understanding of variability. Multilevel modelling with Bayesian methods, therefore, not only improves the prediction of V_s in data-limited regions but also offers a clearer quantification of uncertainty, which is crucial for seismic risk assessment. Despite its potential, this approach is underutilised in geotechnical engineering, highlighting a significant research gap. Addressing this gap could help in the development of N-V_s relationships in sandy soil, making them more reliable and broadly applicable.

The primary aim of this research is to examine and establish the relationship between (V_s) and (N) for sandy soils from various geographical regions. This relationship can be represented by a linear model, $y = a + bx$, where the natural logarithm is applied to both sides of the mentioned equations in literature, transforming Eq. (3.1) into

$$ln\,(Vs) = ln\,(a)\; + b * ln\,(N) \tag{3.2}$$

Here, *ln(a)* is the intercept, and *b* is the slope. Let's analyse the given empirical equations to determine the variability in slopes and intercepts. Empirical formulas are in the form of Eq. (3.1), and by taking the natural logarithm of both sides, the equation becomes in the form of Eq. (3.2).

The relationship between V_s and N across different regions for sandy soils in Table 3.1 is transformed to the logarithmic form line Eq. (3.2) and is shown in Fig. 3.1. It reveals important insights into the variability of slopes and intercepts. While there is some variability, most slopes cluster between 0.265 and 0.5535. The intercepts *ln(a)* vary based on the different values of *a*. The empirical results show that the mean slope is 0.362 with a standard deviation of 0.097, indicating minimal variability. In contrast, the intercepts exhibit a mean of 4.44 with a standard deviation of 0.3937, suggesting significant regional differences in baseline conditions as seen in Fig. 3.1. These findings support our decision to use a multilevel model with varying intercepts but fixed slopes. The small variation in slopes across regions supports our hypothesis, treating the slope as constant, ensuring a parsimonious and interpretable model. Gelman (2007) and Snijders and Bosker (2011) suggest this approach while operating in multi-level modelling. This approach aligns with the principle of Occam's Razor, favouring simplicity without sacrificing explanatory power.

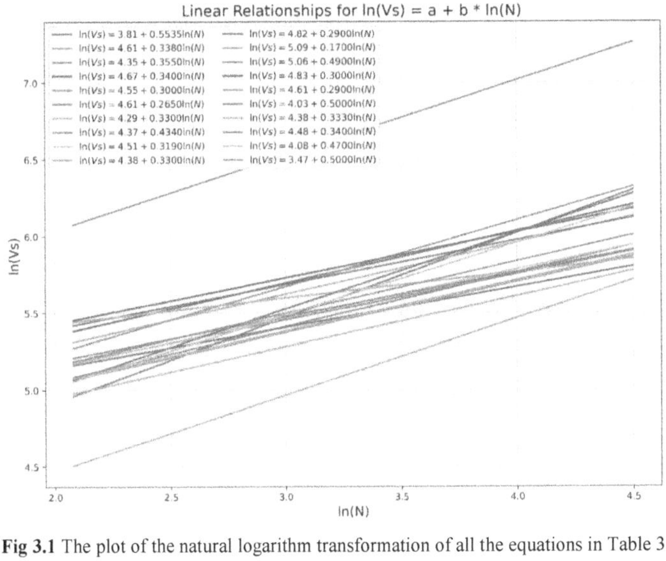

Fig 3.1 The plot of the natural logarithm transformation of all the equations in Table 3.1

Theoretically, the consistent physical processes across regions underpin the relationship between $ln(V_s)$ and $ln(N)$ in sandy soil, while regional differences are more likely to affect the baseline conditions. By modelling the intercepts as varying, we capture these regional differences effectively. Practically, a model with varying intercepts and fixed slopes may provide clear and reliable insights for practitioners and stakeholders. It enhances the robustness and generalizability of our findings, ensuring that the model remains applicable across different regions with distinct baseline conditions.

3.2.1 Data

Our study will focus on the development of the *N*-V_s relationship in sandy soil using the uncorrected SPT-*N* value. The use of uncorrected SPT-N values in this study stems from both practical and methodological considerations. Research has consistently suggested that V_s predictions based on uncorrected SPT-N values often exhibit greater accuracy compared to those derived from energy-corrected values (Kirar et al. 2016; Piratheepan 2002; Hasancebi and Ulusay 2007; and Dikmen 2009). Moreover, the potential influence of geologic age and soil type on N-V_s relationships remains a topic of debate. Several studies, including those by Hasancebi and Ulusay (2007), Maheswari et al. (2010), and Thokchom et al. (2017), have suggested that these factors may have a negligible impact, thereby justifying their exclusion from certain predictive models. Xiao et al. (2021) developed N-Vs relationships for general soil types, which are applicable to regions without region-specific data and regions using the uncorrected SPT-N values. The database for sandy soil is derived from SOIL/2/2433 compiled by Xiao et at (2021) and has 1071 data points. They obtained data that are often presented in graphs; these graphs were digitised using the software *GetData*, which results in a calibration database from different regions worldwide. Table 3.2 summarizes the details of datasets collected from the literature (N_{avg} – Average of SPT-N values, Vs_{avg}- Average of Vs values, *N(std)*- Standard deviation in N values, *Vs(std)*- standard deviation in Vs values) *and* a scatter plot is presented in Fig 3.2 and a boxplot in supplement (Fig 3.S1a, b). It can be observed in Table 3.2 that some N-values exceed the standard upper limit of 50. Many *N-Vs* relationships have been developed without excluding data with $N > 50$ (e.g., Brandenberg et al., 2010; Fabbrocino et al., 2015). Xiao et al. (2021) used N>50 and even N>100 to develop a hierarchical model for general soil types. As data with $N > 50$ in this study do not show a different trend from those with $N \leq 50$, data with $N > 50$ were also considered. Anbazhagan et al (2022) interpreted that N-values are highly sensitive to the energy delivered during the SPT. As most of the equipment in the USA delivers approximately 60% of the theoretical maximum energy, variations in energy efficiency among SPT hammers can lead to discrepancies in measured N-values. For example, a lower energy efficiency hammer will produce higher N-values for the same sandy soil compared to a higher energy efficiency hammer. Standardising N-values to a site-specific energy level is essential to reduce variability caused by inconsistent energy delivery. This ensures that the in-situ soil properties correlated with N-values remain accurate and reliable across different regions and equipment. In India, uncorrected N-values greater than 50 or 100 are often used in sandy rebound layers, as specified in tender documents, and sometimes along with penetration values in weathered rock layers. The Indian standard IS: 2131-1981 (Reaffirmed 2002) recommends a cumulative blow count limit of 50 for the last two increments to terminate boreholes, differing from ASTM D4633-16, which suggests termination when blow counts exceed 50 for one increment or 100 in total. This discrepancy highlights the need to account for energy levels during SPT to avoid errors. Relying solely on arbitrary limits, such as N = 50 or 100, without considering energy corrections or providing complete blow count and penetration details, can result in the misclassification of sandy soil states. Therefore, the overall blow count was not restricted to 50 in this study. Some additional data points from the regions Southern San Francisco Bay, US (Fumal 1978) and Yenisehir town, Turkey (Hasancebi and Ulusay 2007), a combined dataset of 52 points, is used as a validation dataset to check our prediction.

Table 3.2. Details of datasets used

Origin	Region	literature	N	Vs (m/s)	Navg	Vsavg (m/s)	N(std)	Vs(std) (m/s)
1	Madison (US)	Rahimi et al. (2018)	1-30	101-235	13.1	172.8	7.2	31.5
2	Chennai (India)	Maheswari et al. (2010)	8-50	164.86-478.4	27.8	335.1	11.8	76.21
3	Eskisehir (Turkey)	Dikmen (2009)	1-48	100-284.3	24	196.7	12.5	45.1
4	Dholera (India)	Thokchom et al. (2017)	1-49	100-369.2	29.3	212.1	12.5	45.1
5	Caspian Sea (Iran)	Esfehanizadeh et al. (2015)	3-50	118.17-433.2	23.6	311.2	12.1	77.5
6	Roorkee (India)	Kirar et al. (2016)	3-80	100.4-349.5	23.3	216.8	12.1	56.2
7	Erbaa (Turkey)	Akin et al. (2011)	2-18	120.2-285	9.5	211.1	4.2	42.1
8	Molise (Italy)	Fabbrocino et al. (2015)	12-90	196.6-465.4	42.2	303.5	17.7	61.6
9	Kathmandu Valley (Nepal)	Gautam (2017)	8-90	170.2-549.3	41.52	325.6	19.4	90.5
10	Delhi (India)	Hanumanthara o and Ramana (2008)	4-42	90.3-401.2	22.5	233.9	10.5	70.3

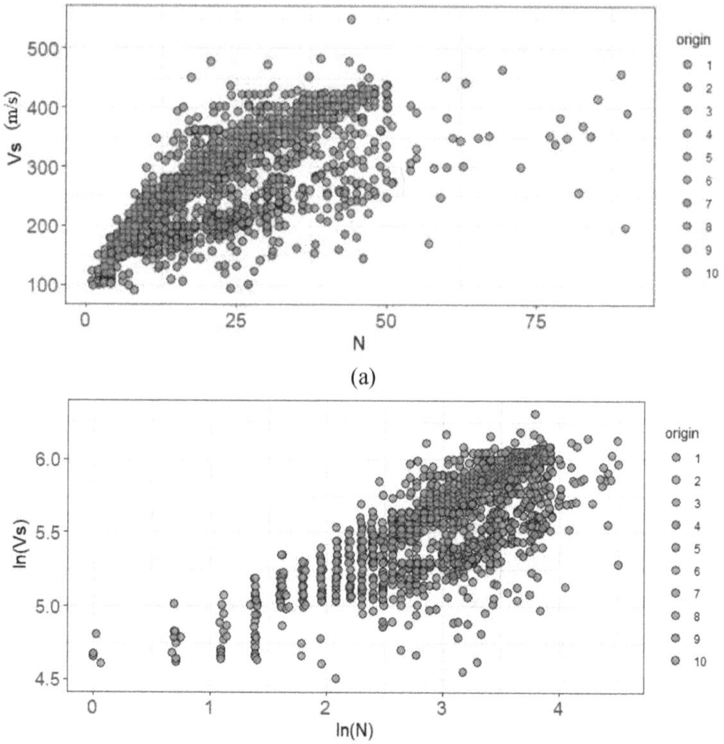

(a)

(b)

Fig 3.2 Scatter plots of V_s *(m/s)* and N data for all 10 origins (a) V_s-N (b) $log(V_s) - log(N)$

3.3 Approach

3.3.1 Statistical analysis

In statistical analysis, there are two primary philosophical approaches: frequentist and Bayesian. Over the past century, the frequentist approach has been predominant. This method bases probabilities on the frequencies of events occurring over repeated trials. Within this framework, parameters are considered fixed and unchanging, while the data, which represent a sample from a broader population, are seen as variable. Frequentist methods include null hypothesis testing, p-values, ANOVA, and confidence intervals. Although effective, interpreting frequentist results can be complex because they relate to the data, not directly to the hypotheses or parameters. This is expressed through the conditional probability $P(D \mid H)$, which is the probability of observing data D given a hypothesis H.

Conversely, the Bayesian approach treats parameters as variables and the data as fixed once observed. Bayesian analysis focuses on the probability of a hypothesis given the data, denoted as $P(H \mid D)$. In this framework, parameters are inferred as probability distributions. For instance, a Bayesian statement might express a 95% probability that a parameter lies within a specific range, which differs from the frequentist approach, where such probability statements about parameters are not possible. A key distinction between the two approaches lies in how they interpret confidence and credible intervals. A frequentist confidence interval is about how frequently a parameter would fall within a certain range over infinite repetitions of an experiment, whereas a Bayesian credible interval directly states the probability that the parameter lies within that range, given the observed data (McElreath 2018).

The Bayesian approach utilises Bayes' theorem, represented by the formula in Eq. (3.3).

$$P(H \mid D) = P(D \mid H) \cdot \frac{P(H)}{P(D)} \tag{3.3}$$

This theorem converts the probability of the data given the hypothesis $P(D|H)$ into the probability of the hypothesis given the data $P(H|D)$, where $P(H)$ is the prior probability and $P(D)$ is a normalisation factor. While Bayesian methods provide a more intuitive interpretation of statistical results by expressing uncertainty through probability distributions, they are computationally intensive. The rise of advanced computing and techniques like Markov Chain Monte Carlo (MCMC) has made it feasible to apply Bayesian methods to complex problems (Gelman and Shalizi 2013). Bayesian methods require specifying a likelihood function and prior probabilities, promoting transparency and debate over assumptions, though some critics argue this subjectivity compromises scientific objectivity.

Variability is an inherent characteristic in the study of geotechnical engineering. Traditional methods often mask this variability by averaging data, which can lead to a loss of valuable information. Instead, explicitly quantifying this variability is crucial, and multilevel modelling provides a robust framework for doing so. The Bayesian approach is particularly well-suited for this purpose because it accommodates variability in parameters. While frequentist approaches also attempt to model variability through likelihood methods, Bayesian methods offer distinct advantages. Modelling serves two primary purposes: to understand the relationships between variables scientifically and to make predictions. It is possible to predict outcomes based on empirical relationships without fully understanding the underlying mechanisms. However, predictions grounded in scientific and mechanistic understanding tend to be more reliable. This chapter focuses on achieving such reliable predictions and assessing the uncertainty associated with them, which is of great practical significance.

3.3.2 Multi-level modelling

In practical sampling, measurements or observations are often correlated. Data from a particular region or soil type share common characteristics. Standard regression methods assume independence among data points, which is often unrealistic. Multilevel modelling, on the other hand, accounts for these dependencies, leading to more accurate and unbiased estimates and providing better insights into variability.

A Multi-level model is used to analyse the data that are organised at more than one level. In Fig. 3.3, the data are clustered into several regions (Region 1, Region 2, ..., Region r, ..., the new region), and within each region, there are multiple observations. The model includes region-specific parameters (a_1, b) and higher-level parameters (μ_a ,σ_a), indicating that the observations within each region are influenced by both local and global factors. This hierarchical structure allows for the modelling of both within-region variability and between-region variability, capturing the complex dependencies in the data.

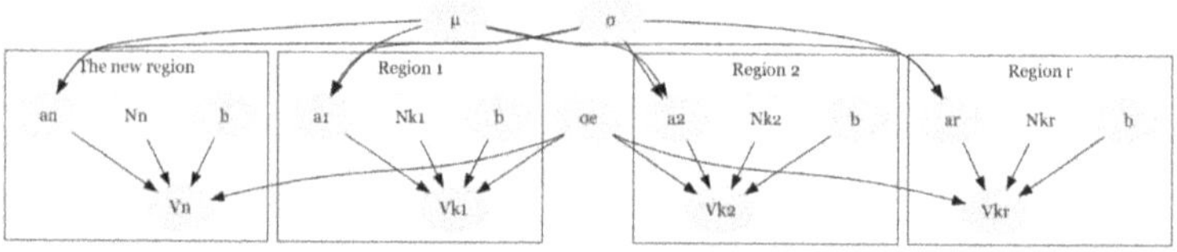

Fig 3.3 A Multi-level model

In terms of clustering, the model effectively groups observations into regions, where each region can have its distinct parameters while also sharing common hyperparameters that govern the overall distribution. Clustering in this multi-level model works by associating each observation (V_{1k}, V_{2k}, etc.) with its respective region-specific parameters (a_r, b), which in turn are influenced by global parameters (μ_a,σ_a). This nested structure allows for borrowing strength across regions, improving parameter estimates for regions with fewer observations by leveraging information from regions with more data. Thus, the multi-level model not only accounts for the hierarchical organisation of the data but also facilitates clustering by allowing for region-specific effects within a common framework.

3.4. Null model

The simplest model for analysing the data is a null model, also known as an intercept-only or unconditional model, which doesn't rely on predictor variables. This model provides estimates for the mean and standard deviation of the population. In this study, a null model can be implemented in three ways: i) pooling all data, ii) analysing each origin cluster separately (no-pooling), and iii) using partial pooling (multilevel modelling). When using Bayesian regression, likelihood and priors need to be defined. The model specifies that the logarithm of the response variable $ln(V_s)$ follows a normal distribution (Zhai et al 2024) since it has been shown that the shear-wave velocity in sandy soil is log-normally distributed (Taylor et al. 2022).

$$\begin{aligned} ln(Vs) &\sim Normal\ (\mu, \sigma) \\ \mu &\sim Normal\ (0, 100) \\ \sigma &\sim exponential(1) \end{aligned} \tag{3.4}$$

This prior information indicates that *ln (Vs)* is modeled as a normally distributed variable with an uncertain mean μ and variability σ in Eq. (3.4). The mean μ has a broad normal prior (0, 100), in reflecting significant initial uncertainty, while the standard deviation σ follows an exponential distribution with a rate of 1, implying smaller values are more probable but allowing for higher variability.

3.4.1 Complete pooling with the null model

In Bayesian regression using a fully pooled method for the parameter *ln(V_s)*, the intercept was estimated to have a mean value of 5.5316 with a standard error of 0.0108, indicating a precise estimate. The standard deviation σ was estimated at 0.3524 with a standard error of 0.0077. Supplementary Fig. 3.S2 provides the complete posterior distribution for this parameter, and diagnostic checks are shown in Fig. 3.S2 to S6. Supplementary Fig. 3.S6 demonstrates that the posterior predictions are closely aligned with the observed data.

3.4.2 No pooling with the null model

Despite employing identical priors and likelihood functions (Eq. (3.4)), separate analyses of origin yielded 10 distinct estimates, highlighting substantial heterogeneity between groups (Fig 3. 4). While complete pooling risks underfitting by neglecting this variation (observed in Fig 3. 2), unconstrained no-pooling suffers from overfitting, treating each dataset in isolation. Partial pooling offers a potential solution, leveraging the strengths of both approaches.

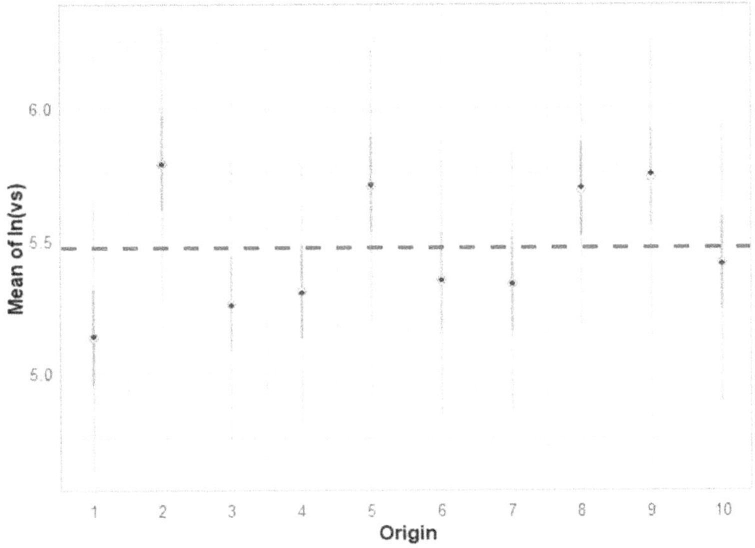

Fig 3.4 Comparison of the mean of *ln(V_s)* obtained through no-pooling (closed black markers) and partial pooling (open green markers with 50% (thick line) and 95% credible interval (thin line)), the dotted horizontal line indicates the overall population mean of 5.472 derived from partial pooling

3.4.3 Partial pooling with the null model

Multilevel modelling incorporates partial pooling to account for both common trends and group-specific variations. It posits that group means (μ_j) deviate from a global mean (μ) with between-group variability captured by σ_b (Eq 3.5). μ_j follows a normal distribution around μ

with group offsets (b_j). Total variance (σ) is partitioned into σ_b (between) and σ_r (within). Prior distributions are specified in Eq. (3.5)

$$\begin{aligned} ln(V_s) &\sim Normal\ (\mu_j, \sigma_r\ ) \\ \mu_j &= \mu + b_j \\ b_j &\sim Normal\ (0, \sigma_b) \\ \mu &\sim Normal\ (0, 100) \\ \sigma_b &\sim exponential(1) \\ \sigma_r &\sim exponential(1) \end{aligned} \tag{3.5}$$

Posterior predictive checks (PPCs) confirm the adequacy of the hierarchical model (refer to Fig. 3. S6 in the Supplement). In hierarchical modelling, shrinkage occurs when parameter estimates are 'pulled' towards the overall mean due to partial pooling. This statistical approach strengthens individual estimates, especially in datasets with limited data points per group. Fig 3.4 visualizes shrinkage. Partial pooling reduces variability in estimates compared to no-pooling, which can be particularly beneficial in small samples. Overall, partial pooling mitigates the effects of limited data and overfitting, leading to more reliable parameter estimates in hierarchical models. Origins 3, 4, 5, 8, 9, 10 show a shrinkage effect, while origins 1, 2, 6, 7 do not exhibit a notable 3 shrinkage effect based on the visual comparison of the markers (the green circle shifting towards the red line from the black dots).

Fig 3.5 provides a detailed graphic illustration comparing the posterior distributions obtained from complete pooling and partial pooling regression models for the two parameters of the null model: the mean and the standard deviation. In Fig 3.5(a), two notable effects of partial pooling are observed: firstly, the mean estimate shifts, and secondly, the uncertainty (width) of this estimate increases significantly. Although this increase in uncertainty may initially appear detrimental, it actually reveals that the complete pooling approach gives a misleading impression of the estimate's uncertainty. This occurs because complete pooling ignores the clustering effect, leading to underfitting and the erroneous assumption of no variation between origins, which is clearly incorrect. In Fig 3.5(b), the impact of partial pooling on the estimates of population standard deviations is evident. Partial pooling partitions the total variance into between-cluster variance and residual variance within clusters, resulting in a substantial reduction of the level-1 variance while highlighting the significant and wider variance due to clustering, indicating greater uncertainty in this estimate. This illustrates the superiority of the Bayesian approach in visualising parameter behavior through posterior distributions, offering deeper insights than point estimates alone.

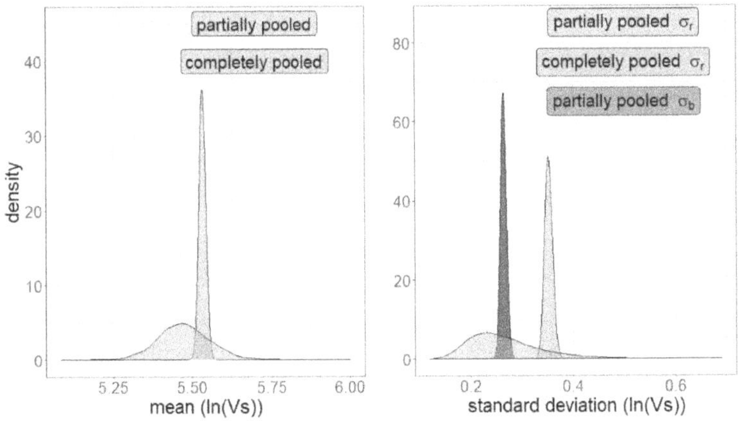

(a) (b)

Fig 3.5 Posterior parameter distributions resulting from pooled and partially pooled regression with the null model (a) means, (b) standard deviations

Table 3.3 provides a summary of the posterior distributions obtained from partial pooling. The parameter σ_b quantifies the variation of individual means around a central value for each origin data set. Detailed numerical offsets from the population mean (parameters b_j) for each origin are presented in Supplementary Fig. 3.S10a. When comparing complete pooling and partial pooling, the intercept estimate (mean) shifts from 5.53 to 5.472, and the standard deviation of 0.3524 decreases due to the variance being partitioned across two levels. Additionally, the intra-class correlation coefficient (ICC), which represents the ratio of between-cluster variance to total variance, can be derived from σ_b and σ_r (refer to Eq. (3.6) for calculation).

$$ICC = \frac{\sigma_b^2}{\sigma_b^2+\sigma_r^2} \tag{3.6}$$

The intra-class correlation coefficient (ICC) of 0.504 in this study indicates that 50.4% of the total variance in $ln(V_s)$ in sandy soil is attributable to differences between clusters (origins), demonstrating a strong dependence on origin. An ICC value of 0 implies no variation between clusters, while an ICC of 1 signifies that all variance is due to differences between clusters. According to Sommet and Morselli (2021), ICC values of 0.01, 0.05, and 0.2 are typically considered to represent small, intermediate, and large degrees of cluster dependence, respectively. The substantial clustering effect observed here highlights the importance of origin in influencing shear velocity, while the remaining 49.6% of the variance, due to within-cluster differences, points to significant variability within each origin. This suggests that while origin is a major factor, other intra-cluster factors also play a crucial role in determining shear velocity. Multilevel modelling is therefore very appropriate.

Table 3 3. Partial pooling results of the $ln(V_s)$- $ln(N)$ data

Parameter	Mean	SE	Lower bound	Upper bound
μ	5.472	0.089	5.307	5.649
σ_b	0.267	0.073	0.164	0.443
σ_r	0.265	0.006	0.254	0.276

3.5. Regression with ln(*N*) as predictor

Centring a predictor variable in regression analysis is achieved by subtracting its mean from each observed value, which offers significant advantages. It renders the intercept more interpretable, representing the expected value of the dependent variable at the mean of the predictor. This technique also reduces multicollinearity, especially when interaction or polynomial terms are present, thereby stabilising coefficient estimates. Additionally, centring decreases the correlation between slope and intercept parameters, enhancing the reliability of parameter estimation (McElreath 2018). It also improves numerical stability and facilitates convergence in iterative estimation algorithms, making it a crucial step for robust and interpretable regression models.

3.5.1 Complete pooling with $ln(N)_c$

Starting with ignoring the clusters in the data and pooling all data, the model proposed is as in Eq. (3.7).

$$ln(V_s) = ln(a) + b * \ln(N)_c \tag{3.7}$$

Again, a likelihood function for the data and priors for the parameters are given in Eq. (3.8)

$$\begin{aligned} ln(V_s) &\sim Normal\ (\mu_i, \sigma_r) \\ \mu_i &= \beta_0 + \beta_1 * ln(N_i)_c \\ \beta_0 &\sim Normal\ (0{,}100) \\ \beta_1 &\sim Normal\ (0, 100) \\ \sigma_r &\sim exponential(1) \end{aligned} \tag{3.8}$$

To model the relationship between the variables, we define a likelihood function and specify priors for the parameters. The likelihood function is given by Eq. (3.8), indicating that the natural logarithm of V_s follows a normal distribution with mean μ_i and standard deviation σ_r. The mean μ_i is expressed as a linear function of the centred predictor $ln(N)_c$, assign prior distributions to the parameters: β_0~Normal (0,100) and β_1~Normal (0,100), which reflect weakly informative priors, allowing the data to primarily inform the parameter estimates. The standard deviation σ_r is assigned an exponential prior, suitable for capturing the positive scale of the residuals. This Bayesian framework integrates prior knowledge and observed data to yield robust parameter estimates. A numerical summary of the posterior is shown in Table 3.4.

Supplement Fig 3.S7 compares posterior distributions, pair plots, and correlation coefficients for regression with non-centred and centred predictor variables. It shows clearly that centring indeed removes a strong parameter correlation between slope and intercept. Therefore, for the remainder of the article, centred $ln(N)_c$ will be used as a predictor. The investigation into the presence of Simpson's paradox in Fig.3.6 reveals intriguing disparities between Fig.3 6(a), encompassing data from all 10 origins, and Fig. 3.6(b), featuring a subset of 4 origins. In Fig. 3.6(a), a coherent positive correlation between $ln(V_s)$ and centred $ln(N)$ emerges, reflecting the combined effect of all origins. However, Fig 3. 6(b) portrays a more nuanced narrative, where the subset's data points exhibit diverse patterns, potentially deviating from the overarching trend observed in Fig 3.6(a). This divergence suggests a complex interplay among the origins, culminating in contrasting relationships between $ln(V_s)$ and centred $ln(N)$ across the two datasets. Consequently, scrutiny of subgroup analyses is imperative to discern the underlying mechanisms driving these discrepancies and ascertain the true nature of the observed trends.

Table 3 4. Results of complete pooling

Parameter	Mean	SE	Lower bound	Upper bound
Intercept β_0	5.532	0.007	5.52	5.55
Slope β_1	0.267	0.010	0.164	0.37
σ_r	0.242	0.005	0.23	0.25

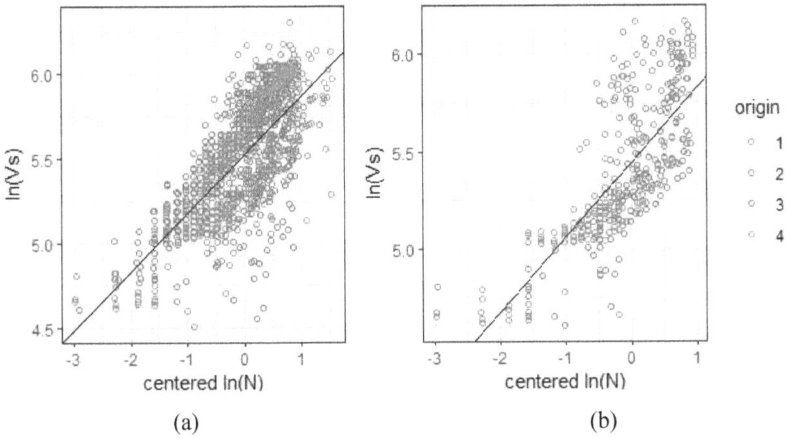

Fig 3.6 Regression plot of (a) complete pooling of data and (b) a subset of origin (1–4) to show the effect of Simpson's paradox using centred $ln(N)_c$ as the predictor

3.5.2 No pooling

No-pooling regression, conducted with identical priors and the likelihood as in complete pooling Eq. (8), diverges in its approach by assigning unique slopes and intercepts to each of 10 clusters. Whereas complete pooling fails to recognise the clustered nature of the data, no-pooling overlooks their shared characteristics, as discussed previously regarding the null model. This results in 10 distinct slopes and intercepts, depicted in dots-and-whisker plots in Supplement Fig 3.S8(a, b), and corresponding fits are shown in Fig 3.7. Although individual fits exhibit acceptable performance, significant differences arise, with notable deviations observed in regression results from origins 2, 4, 8 and 9,10, where the slope for origin 9 even displays negativity (Panel 9's slope is nearly zero or slightly negative because the data there do not show a consistent positive trend, which flattens the regression line and greatly widens the prediction intervals. This illustrates how, without pooling, groups with limited or noisy data can produce unreliable slope estimates and inflated uncertainty). Importantly, the inability to generalise beyond the sampled cases and the wide variance in predictive capacity for individual instances, as evident in Fig. 3.7, underscores the limitations of the no-pooling approach.

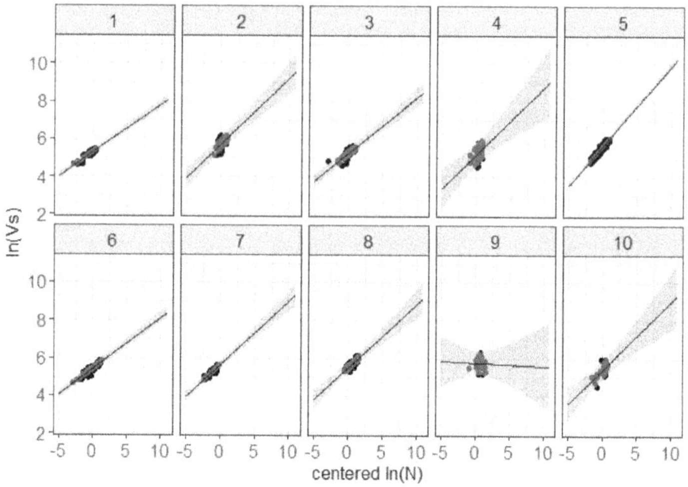

Fig 3.7 Regression lines resulting from individual regressions (no pooling) of the data according to origin using $ln(N)_c$ as a predictor (the grey ribbons represent the 95% prediction intervals)

3.5.3 Partial pooling

Based on the analysis presented in the 'Introduction', it is apparent that the observed variation is predominantly in the intercepts rather than the slopes, as indicated by the standard deviation from the mean of literature reviews and the results discussed in the null model. This suggests that while the intercepts show considerable deviation, the slopes remain relatively consistent across the studies examined.

Multilevel regression is attempted with varying intercepts only. This requires different priors and the likelihood as shown in Eq. (3.9). b_{0j} is the offset for each origin j from the global intercept β_0, while β_1 is the global slope shared by all origins. σ_r describes the within-origin variation and σ_b the between-origin variation.

$$\begin{aligned} ln(Vs) &\sim Normal\left(\mu_{ij}, \sigma_r\right) \\ \mu_{ij} &= \left(\beta_0 + b_{0j}\right) + \beta_1 * ln(N_{ij})_c \\ b_{0j} &\sim Normal\,(0, \sigma_b) \\ \beta_0 &\sim Normal\,(0, 100) \\ \beta_1 &\sim Normal\,(0, 100) \end{aligned} \tag{3.9}$$

$$\sigma_b \sim exponential(1)$$

$$\sigma_r \sim exponential(1))$$

When the model is run and the checks are done on the posterior obtained are found to be ok in Supplementary Fig 3. S9(a, b). The comparison of complete pooling and partial pooling of data can be made for a model with $ln(N)_c$ as a predictor using the posterior distributions. Figure 3.8 (a) shows the distributions for β_0, with the partially pooled model exhibiting wider variability compared to the completely pooled model. Fig 3.8(b) illustrates the distributions for β_1, where the partially pooled model again shows a broader distribution than the completely pooled model. Fig 3. 8(c) presents the distributions for the variance parameters σ_r and σ_b, indicating that the completely pooled σ_r is more concentrated around lower values, while the partially pooled σ_b and σ_r distributions are wider, reflecting greater uncertainty (width) of this estimate. Although this increase in uncertainty may initially appear detrimental, it reveals that the complete pooling approach gives a misleading impression of the estimate's uncertainty. This is also showcased in Table 3 S4 and S5. The difference in lower and upper values is greater in Table 3.5. representing partial pooling, then in Table 3.4, representing complete pooling of data. The advantage of partial pooling is that it balances between group-specific estimates and the overall population estimate, reducing variability and preventing overfitting. Partial pooling shrinks group-level estimates towards the overall mean, improving parameter estimation and capturing variability more realistically. It separates within-group and between-group variance components, leading to a better understanding of hierarchical variance. This approach results in more robust and generalizable models by incorporating both individual and overall data information.

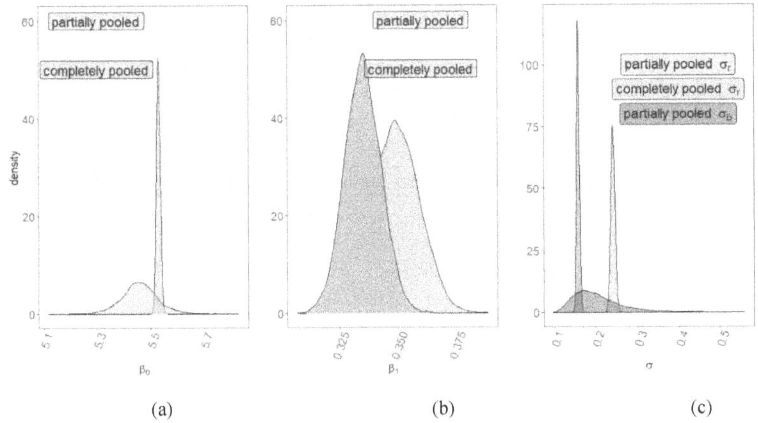

Fig 3.8 Comparison of posterior parameter distributions resulting from regression with centred predictor for complete pooling and partial pooling (a) intercept, (b) slope, (c) standard deviations

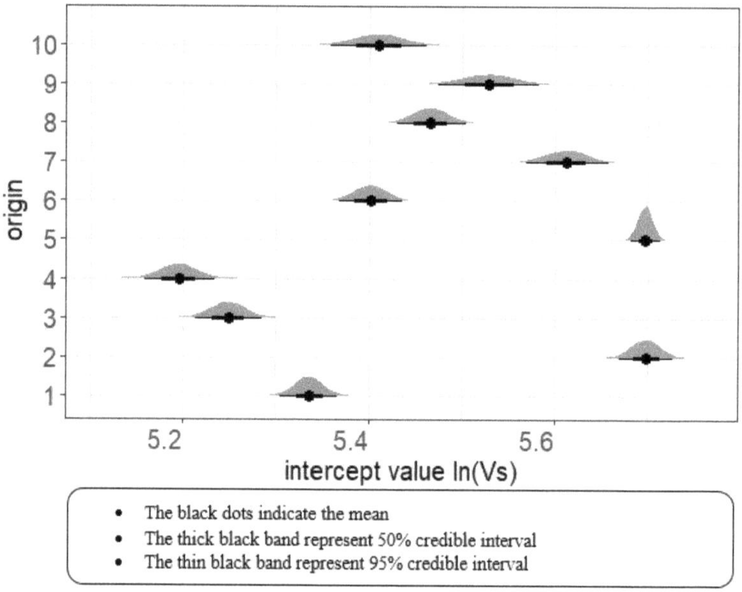

Fig 3.9 The coefficient plot of intercepts for each origin obtained from partial pooling with centred *ln(N)* as the predictor variable

The varying intercepts resulting from partial pooling are shown in Fig. 3.9, and they show the posterior distributions of the intercepts for each site, indicating the variability and uncertainty of these estimates. In Fig. 3.10, regression lines are presented for ten individual datasets, demonstrating the differences between group-level fits from partial pooling, population-level fits from partial pooling, and individual regressions from no pooling. The solid blue lines depict group-level regression fits derived from the partially pooled model, which estimates the relationship between *ln(N)* and $ln(V_s)$ by integrating data from all groups. This approach balances individual dataset trends with the overall population trend, thus capturing within-group variability while leveraging the entire dataset's information. The dashed red lines represent population-level fits from the partially pooled model, which is identical across all panels, indicating a global trend that assumes a common relationship for all data points without considering significant variation among groups. These lines highlight the overarching pattern in the data, providing a baseline for comparison. The black dotted lines show individual regression fits from the no-pooling approach, reflecting the relationship between $ln(Nc)_c$ and $ln(V_s)$ for each dataset separately. These fits capture group-specific trends without borrowing information from other groups, resulting in higher variability and susceptibility to noise and overfitting, particularly in datasets with limited data points or high variance.

The Fig 3.10 contrasts the behaviours of the three approaches. The group-level fit (solid blue line) demonstrates greater stability and less variability compared to the individual regressions (black dotted lines), highlighting the advantage of partial pooling in reducing overfitting and incorporating comprehensive dataset information. The population-level fit (dashed red lines), while providing a useful reference, may oversimplify the data by ignoring group-specific variations. This comparative analysis underscores the advantages of partial pooling in hierarchical models. By finding a balance between the extremes of complete pooling (population-level fit) and no pooling (individual regressions), partial pooling offers a robust modelling approach that enhances parameter estimation and generalizability. It effectively captures both global trends and group-specific nuances, resulting in more reliable and interpretable outcomes. Thus, Fig. 3.10 exemplifies how partial pooling mitigates the risk of overfitting and improves model robustness and accuracy by integrating individual and overall data patterns.

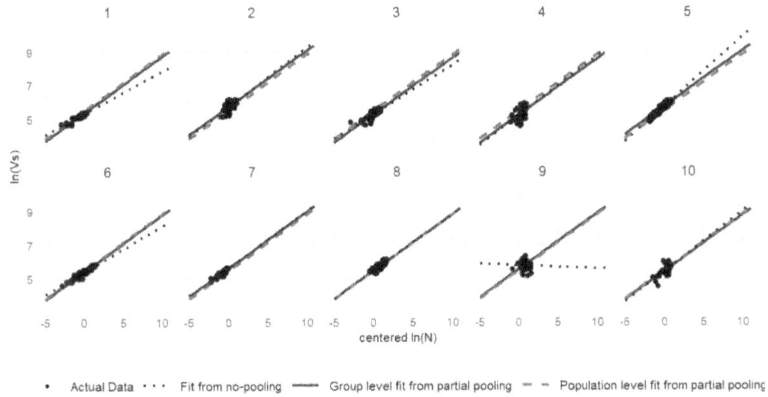

Fig 3.10 Regression lines for individual data sets

Notably, cases 1,5,6,9 in Fig. 3.10 reveal the advantage of partial pooling as the group-level fits (solid blue lines). The partial pooling results for these cases follow the general trend of the other data sets as compared to the extreme variability seen in the individual regressions (black dotted lines), thereby reducing the influence of outliers and noise. In Figures 7 and 10, the horizontal axis represents the centred natural logarithm of N($\ln(N)_c$). Centring subtracts the mean $\ln(N)_c$ from each observed value. $\ln(N)_c$ can be negative when the observed values of $\ln(N)_c$ are less than the mean.

The numerical summary of the posterior distribution resulting from partial pooling is shown in Table 3.5. The actual offsets b_{0j} from the global parameter β_0 for each origin can be found in Supplement Fig. 3. S10(b). The results demonstrate a notable difference between the complete pooling and partial pooling approaches. In the complete pooling model, summarised in Table 3.3, the population estimate for the intercept was 5.532, whereas in the partial pooling model, it shifted to 5.459. This shift in the intercept estimate is indicative of the influence of partial pooling on the model parameters.

Table 3 5. Partial pooling results

Parameter	Mean	SE	Lower bound	Upper bound
Intercept β_0	5.459	0.070	5.318	5.601
Slope β_1	0.335	0.008	0.320	0.350
σ_b	0.202	0.057	0.125	0.346
σ_r	0.158	0.003	0.152	0.165

Furthermore, the parameter σ_r, which represents the residual variance, decreased from 0.242 in the complete pooling model to 0.158 in the partial pooling model. This decrease can be attributed to the partitioning of variance towards σ_b, the between-cluster variance component. In other words, the partial pooling model attributes a larger proportion of the overall variance to the differences between clusters, rather than the residual variance within clusters. This phenomenon is further supported by the intra-class correlation coefficient (ICC) parameter, which now has two distinct values in the partial pooling model. The adjusted ICC value of 0.620 accounts solely for the varying effects, while the unadjusted ICC value of 0.323 incorporates both the varying and population effects. Both ICC values indicate a strong effect of variance partitioning due to the clustering of the data, reinforcing the appropriateness of employing multilevel modelling for this analysis.

3.6 Prediction

In multilevel modelling, prediction depends on the intended goal. To predict at the population level, population parameter estimates or marginal effects can be used, which assess the impact of a variable (N) on the response variable (V_s in sandy soil) across clusters. For predictions at existing cluster levels, the group-specific parameter estimates can be utilised. Additionally, predictions for new, uninvestigated clusters can be derived from equation 10. The random effect b_{0j} is unknown, but its prior distribution is used, and the prediction incorporates both b_{0j} and residual uncertainty (σ_r):

$$b_{0j} \sim Normal(0, \sigma_b) \tag{3.10}$$

$$ln(V_s)_{ij}^{new} \sim Normal\left((\beta_0) + \beta_1 * \ln(N_{ij})_c, \sqrt{\sigma_b{}^2 + \sigma_r{}^2}\right)$$

$$ln(V_s)_{ij}^{new} \sim Normal\left((5.459) + 0.335 * \ln(N_{ij})_c, \sqrt{0.202^2 + 0.158^2}\right)$$

β_0, β_1, σ_b, σ_r values are obtained from Table 3 5. This approach in multilevel modelling helps in mitigating overfitting and underfitting, ensuring more robust and generalizable predictions (McElreath 2018).

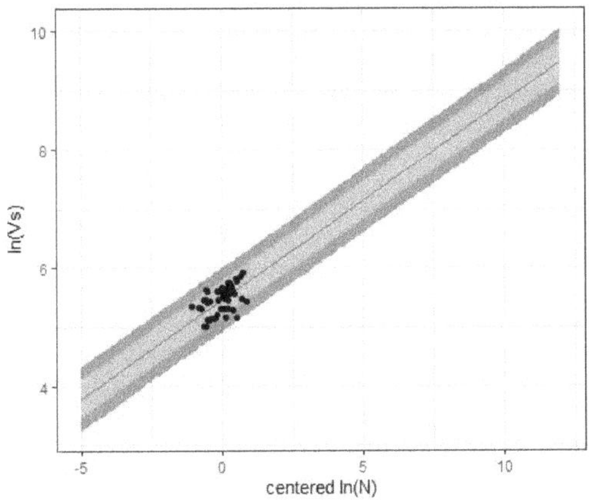

Fig 3.11 The yellow band shows the 95% prediction interval for origin 8, the red band for an unknown origin, and the black points are the unknown-origin data.

Fig 3.11. illustrates the effectiveness of a multilevel regression model in predicting outcomes for both known and unknown origins, with a particular emphasis on evaluating the fit with data points from an existing origin (labelled as Origin 8) and new, previously unsampled origins (origin 8 was used for illustration because its individual regression line coincides with the population regression line). The yellow ribbon on the plot delineates the 95% prediction interval for new samples originating from existing origins, specifically including Origin 8. This narrower prediction interval signifies a higher degree of confidence in the model's predictions for this origin.

Conversely, red ribbon on the plot indicates the 95% prediction interval for samples from new, unknown origins (black dots). This interval is notably wider, reflecting the increased uncertainty inherent in making predictions for groups that were not part of the model's building data. Despite this broader range, the prediction interval remains informative, offering a plausible range within which new observations are likely to fall. The distinction between the prediction intervals for existing and unknown origins highlights the multilevel model's strength in handling hierarchical and nested data structures. By accounting for the variability both within and between groups, the model provides robust predictions that incorporate the uncertainty associated with each level. The good fit with the observed data points for Origin 8, as well as the reasonable predictions for unknown origins, underscores the model's flexibility and reliability in diverse research contexts. The thing to remember is that the model primarily captures the population trend, and individual data points, especially from unknown origins, might deviate more significantly. Multilevel models offer a unique advantage - predicting behaviour for entirely new clusters. Unlike existing samples with specific details, new clusters lack individual data. However, multilevel modelling can leverage general population information and variation around the global mean to make estimates. This is achieved by simulating from the posterior distribution, which incorporates three key sources of variability: 1) within-group sampling variation, 2) between-group sampling variation, and 3) variability in the global parameters themselves.

3.7 Conclusion

The integration of multilevel modelling (MLM) has facilitated the development and calibration of a linear model elucidating the relationship between the natural logarithm of shear wave velocity ($ln(V_s)$) and the natural logarithm of the number of blow counts ($ln(N)$) in sandy soils. The foundational hypothesis, based on an extensive review of the V_s-N relationship literature, posits that the transformation to natural logarithms produces a linear relationship with a relatively stable slope, while the intercept varies. This variation in intercepts is attributable to the inherent uncertainties in data sourced from diverse geographical regions. The primary goal of this model is to provide a straightforward yet effective tool for both researchers and field engineers, where the slope remains constant across different locations and the intercepts reflect region-specific influences and data collection uncertainties.

By employing MLM, the variability in the intercepts is quantified, offering a nuanced understanding of regional data variability. MLM's ability to account for data interdependence and provide realistic variability estimates through partial pooling distinguishes it from single-level models, which risk overfitting or underfitting. Partial pooling combines data from various regions, establishing a general trend while accommodating regional deviations, thereby enhancing predictive accuracy and generalizability across contexts. The Bayesian approach further augments MLM by using probability distributions to represent parameter behaviour, thereby quantifying uncertainties and improving predictive capacity. Consequently, the combination of MLM and Bayesian methods culminates in a robust and calibrated predictive model for the $ln(V_s)$-$ln(N)$ relationship in sandy soils, adaptable to other soil types, thus demonstrating significant potential for broader applications in geotechnical engineering research.

Reference

Adamo, F., et al. "Soil moisture assessment by means of compressional and shear wave velocities: Theoretical analysis and experimental setup." *Measurement* 43.3 (2010): 344-352. https://doi.org/10.1016/j.measurement.2009.11.007.

Akin, Muge K., Steven L. Kramer, and Tamer Topal. "Empirical correlations of shear wave velocity (Vs) and penetration resistance (SPT-N) for different soils in an earthquake-prone area (Erbaa-Turkey)." *Engineering geology* 119.1-2 (2011): 1-17.https://doi.org/10.1016/j.enggeo.2011.01.007

Anbazhagan, Panjamani, et al. "Effective use of SPT: Hammer energy measurement and integrated subsurface investigation." *Indian Geotechnical Journal* 52.5 (2022): 1079-1096.https://doi.org/10.1007/s40098-022-00609-z.

ASTM D4633 Standard test method for energy measurement for dynamic penetrometers. ASTM International, West Conshohocken, PA (2016). https://doi.org/10.1520/d4633-16.

Brandenberg, Scott J., Naresh Bellana, and Thomas Shantz. "Shear wave velocity as a function of standard penetration test resistance and vertical effective stress at California bridge sites." *Soil Dynamics and Earthquake Engineering* 30.10 (2010): 1026-1035. https://doi.org/10.1016/j. soildyn.2010.04.014.

Daag, Arturo S., et al. "Empirical correlation between standard penetration resistance (SPT-N) and shear wave velocity (Vs) for soils in Metro Manila, Philippines." *Applied Sciences* 12.16 (2022): 8067.https://doi.org/10.3390/app12168067

Dikmen, Ünal. "Statistical correlations of shear wave velocity and penetration resistance for soils." *Journal of Geophysics and Engineering* 6.1 (2009): 61-72.https://doi.org/10.1088/1742-2132/6/1/007

Esfehanizadeh, Mehran, Farhad Nabizadeh, and Rasool Yazarloo. "Correlation between standard penetration (N SPT) and shear wave velocity (VS) for young coastal sands of the Caspian Sea." *Arabian Journal of Geosciences* 8 (2015): 7333-7341. https://doi.org/10.1007/s12517-014-1751-x

Fabbrocino, Silvia, et al. "SPT blow count vs. shear wave velocity relationship in the structurally complex formations of the Molise Region (Italy)." *Engineering Geology* 187 (2015): 84-97.https://doi.org/10.1016/j.enggeo.2014.12.016.

Fatehnia, Milad, and Gholamreza Amirinia. "A review of genetic programming and artificial neural network applications in pile foundations." *International Journal of Geo-Engineering* 9.1 (2018): 2.https://doi.org/10.1186/s40703-017-0067-6

Fooladi, H., A. Johari, and M. Behzadi. "Comparative System Reliability Analysis of Liquefaction Potential Using Standard Penetration Test and Shear Wave Velocity Data." *Geotechnical and Geological Engineering* 42.8 (2024): 7635-7653.https://doi.org/10.1007/s10706-024-02943-w.

Fumal, Thomas E. *Correlations between seismic wave velocities and physical properties of near-surface geologic materials in the southern San Francisco Bay region, California*. No. 78-1067. US Geological Survey, 1978.

Gautam, Dipendra. "Empirical correlation between uncorrected standard penetration resistance (N) and shear wave velocity (VS) for Kathmandu Valley, Nepal." *Geomatics, Natural Hazards and Risk* 8.2 (2017): 496-508.https://doi.org/10.1080/19475705.2016.1243588

Gelman, Andrew. "Comment: Bayesian checking of the second levels of hierarchical models." *Statistical science* 22.3 (2007): 349-352.

Gelman, Andrew, and Cosma Rohilla Shalizi. "Philosophy and the practice of Bayesian statistics." *British Journal of Mathematical and Statistical Psychology* 66.1 (2013): 8-38.

Hanumantharao, C., and G. V. Ramana. "Dynamic soil properties for microzonation of Delhi, India." *Journal of earth system science* 117 (2008): 719-730.https://doi.org/10.1007/s12040-008-0066-2

Hasancebi, Nilsun, and Resat Ulusay. "Empirical correlations between shear wave velocity and penetration resistance for ground shaking assessments." *Bulletin of Engineering Geology and the Environment* 66 (2007): 203-213.https://doi.org/10.1007/s10064-006-0063-0.

Huang, Zhong-Kai, and Dong-Mei Zhang. "Scalar-and vector-valued vulnerability analysis of shallow circular tunnel in soft soil." *Transportation Geotechnics* 27 (2021): 100505.https://doi.org/10.1016/j.trgeo.2020.10050.

IS 2131 Indian standard method for standard penetration test for soils (first revision) reaffirmed 2002. Bureau of Indian Standards, New Delhi (1981).

Jafari MK, Shafiei A, Razmkhah A Dynamic properties of fine grained soils in south of Tehran (2002).

Japan Road Association. "Specification and interpretation of bridge design for highway—Part V: resilient design." (1980): 14-15.

Johari, A., A. R. Khodaparast, and A. A. Javadi. "An analytical approach to probabilistic modeling of liquefaction based on shear wave velocity." *Iranian Journal of Science and Technology, Transactions of Civil Engineering* 43 (2019): 263-275.https://doi.org/10.1007/s40996-018-0163-7.

Kanai, K., et al. "Observation of microtremors, XI: Matsushiro earthquake swarm areas." *Bull. Earthq. Res. Inst* 44.Part 3 (1966).

Kirar, Bablu, Bal Krishna Maheshwari, and Pradeep Muley. "Correlation between shear wave velocity (Vs) and SPT resistance (N) for Roorkee region." *International Journal of Geosynthetics and Ground Engineering* 2 (2016): 1-11.https://doi.org/10.1007/s40891-016-0047-47

Koçkar, Mustafa K., and Haluk Akgün. "Development of a geotechnical and geophysical database for seismic zonation of the Ankara Basin, Turkey." *Environmental geology* 55 (2008): 165-176.https://doi.org/10.1007/s00254-007-0973-6

Lee, Shannon Hsien-Heng. "Analysis of the multicollinearity of regression equations of shear wave velocities." *Soils and foundations* 32.1 (1992): 205-214.

Maheshwari, Bal Krishna, et al. "Relationship between shear velocity and SPT resistance for sandy soils in the Ganga basin." *International Journal of Geotechnical Engineering* 7.1 (2013): 63-70. https://doi.org/10.1179/1938636212Z.0000000007

Uma Maheswari, R., A. Boominathan, and G. R. Dodagoudar. "Use of surface waves in statistical correlations of shear wave velocity and penetration resistance of Chennai soils." *Geotechnical and Geological Engineering* 28 (2010): 119-137.https://doi.org/10.1007/s10706-009-9285-9

McElreath, Richard. *Statistical rethinking: A Bayesian course with examples in R and Stan*. Chapman and Hall/CRC, 2018.

Ohsaki, Yorihiko, and Ryoji Iwasaki. "On dynamic shear moduli and Poisson's ratios of soil deposits." *Soils and Foundations* 13.4 (1973): 61-73.

Ohta, Yutaka, and Noritoshi Goto. "Empirical shear wave velocity equations in terms of characteristic soil indexes." *Earthquake engineering & structural dynamics* 6.2 (1978): 167-187.**https://doi.org/10.1002/eqe.4290060205**

Okamoto, T., et al. "Comparison of surface versus subsurface wave source for P–S logging in sand layer." *Proc. 44th Ann. Conf. JSCE*. Vol. 3. 1989.

Oliveira, Liliana, et al. "Seismic microzonation based on large geotechnical database: Application to Lisbon." *Engineering Geology* 265 (2020): 105417. https://doi. org/10.1016/j.enggeo.2019.105417.

Piratheepan, Paramananthan. "Estimating shear-wave velocity from SPT and CPT data." (2002).

Pitilakis, K. D., A. Anastasiadis, and D. Raptakis. "Field and laboratory determination of dynamic properties of natural soil deposits." *Proceedings of the 10th world conference on earthquake engineering*. Vol. 5. 1992.

Rahimi, Salman, et al. "The combined use of MASW and resistivity surveys for levee assessment: A case study of the Melvin Price Reach of the Wood River Levee." *Engineering Geology* 241 (2018): 11-24.https://doi.org/10.1016/j.enggeo.2018.05.009

Raptakis, D. G., et al. "Shear wave velocities and damping of Greek natural soils." *Proc. 10th European Conf. Earthquake Engg., Vienna*. Vol. 477482. 1995. https://doi.org/10.1061/(ASCE)1090-0241(1998)124:5(396).

Seed, H. Bolton, Izzat M. Idriss, and Ignacio Arango. "Evaluation of liquefaction potential using field performance data." *Journal of geotechnical engineering* 109.3 (1983): 458-482. https://doi.org/10.1061/(ASCE)0733-9410(1983)109:3(458)

Bolton Seed, H., et al. "Influence of SPT procedures in soil liquefaction resistance evaluations." *Journal of geotechnical engineering* 111.12 (1985): 1425-1445.https://doi.org/10.1061/(ASCE)0733-9410(1985)111:12(1425).

Shibata, Toru, Hiroshi Yukitomo, and Manabu Miyoshi. "Liquefaction process of sand during cyclic loading." *Soils and Foundations* 12.1 (1972): 1-16.

Snijders, Tom AB, and Roel Bosker. "Multilevel analysis: An introduction to basic and advanced multilevel modeling." (2011): 1-368.

Sommet, Nicolas, and Davide Morselli. "Keep calm and learn multilevel linear modeling: A three-step procedure using SPSS, Stata, R, and Mplus." *International Review of Social Psychology* 34.1 (2021).

Sykora, David Wayne. *Correlations of in situ measurements in sands of shear wave velocity, soil characteristics, and site conditions*. Diss. University of Texas at Austin, 1983.

Taylor, Oliver-Denzil S., Masood Abdollahi, and Farshid Vahedifard. "Statistical distributions of wave velocities and elastic moduli in near-surface unsaturated soils." *Soil Dynamics and Earthquake Engineering* 157 (2022): 107247. https://doi.org/10.1016/j.soildyn.2022.107247.

Thokchom, Sarda, et al. "Empirical correlation of SPT blow counts versus shear wave velocity for different types of soils in Dholera, Western India." *Natural Hazards* 86 (2017): 1291-1306.https://doi.org/10.1007/s11069-017-2744-3.

Tong, Liyuan, et al. "Review of shear-wave velocity prediction equations based on piezocone penetration test data: example from Yangtze River floodplain deposits at Nanjing, Jiangsu Province, China." *Quarterly Journal of Engineering Geology and Hydrogeology* 51.2 (2018): 229-246.https://doi.org/10.1144/qjegh2016-088.

Tsiambaos, G., and N. Sabatakakis. "Empirical estimation of shear wave velocity from in situ tests on soil formations in Greece." *Bulletin of Engineering Geology and the Environment* 70 (2011): 291-297.https://doi.org/10.1007/s10064-010-0321-9.

Wair, Bernard R., Jason T. DeJong, and Thomas Shantz. *Guidelines for estimation of shear wave velocity profiles.* Pacific Earthquake Engineering Research Center, 2012.

Xiao, Shihao, et al. "Establishing region-specific N–Vs relationships through hierarchical Bayesian modelling." *Engineering Geology* 287 (2021): 106105.https://doi.org/10.1016/j.enggeo.2021.106105

Zhai, Shijie, Guangyin Du, and Huan He. "Bayesian probabilistic characterisation of the shear-wave velocity combining the cone penetration test and standard penetration test." *Stochastic Environmental Research and Risk Assessment* 38.1 (2024): 69-84. https://doi.org/10.1007/s00477-023-02566-2.

Zhang, Dong-Mei, et al. "An incident of water and soil gushing in a metro tunnel due to high water pressure in sandy silt." *Engineering Failure Analysis* 121 (2021): 105196. https://doi.org/10.1016/j.engfailanal.2020.105196.

Zhang, J., et al. "Calibrating cross-site variability for reliability-based design of pile foundations." *Computers and Geotechnics* 62 (2014): 154-163. https:// doi.org/10.1016/j.compgeo.2014.07.013.

CHAPTER

4

Bayesian-Optimised Ensemble Models for Spread Footings on Clay

4.1 Introduction

In geotechnical engineering, spread footings are primarily designed to meet either the ultimate limit state (ULS) or the serviceability limit state (SLS). At ULS, the focus is on the bearing capacity, which ensures the foundation can withstand maximum applied loads without failure. On the other hand, SLS accounts for the allowable settlement, ensuring that deformations remain within permissible limits to prevent structural or functional impairments (Omar et al. 2018). Traditionally, the determination of ULS bearing capacity relies on empirical methods, but these approaches often fail to account for the variability and complexity inherent in real-world soil-structure interactions (Das and Sivakugan 2018). Experimental research in geotechnical engineering often involves the use of smaller-scale models that replicate actual foundations in a controlled laboratory environment. These scale models are significantly reduced in size compared to full-scale structures, enabling researchers to conduct tests under manageable and replicable conditions (Khajehzadeh et al. 2022a). While these scaled experiments provide valuable insights into the behaviour of foundations, their applicability to real-world scenarios is inherently limited by the challenges of scale effects, which may introduce discrepancies when translating laboratory findings to full-scale applications (Padmini et al. 2008).

The quest for more reliable and precise methods to predict bearing pressure has led researchers to leverage load test data. Both real-size foundations and scaled-down laboratory models have been employed for this purpose, providing valuable insights into soil behaviour under loading conditions (Khajehzadeh and Keawsawasvong 2024). Among these, the plate load test (PLT) has emerged as a dependable technique. It serves a dual role by providing data for both ULS and SLS evaluations, thus becoming a cornerstone in footing design practices. Plate load test remains the most dependable method for directly determining the bearing pressure (*P*) of foundations. However, practical constraints, including financial limitations and time restrictions, frequently hinder the widespread adoption of such field tests. Despite their utility, the challenge lies in bridging the gap between these controlled experimental conditions and the complex, variable nature of field conditions encountered in practice (De Beer 1965; Yamaguchi 1977). The bearing pressure corresponding to a specific settlement is influenced by various factors and requires detailed investigation to accurately determine its behaviour. In the case of soft soils, numerical analysis is often the preferred approach. This is because collecting undisturbed samples from soft soils is challenging, and any disturbance can significantly alter the soil's properties, leading to inaccurate results (Wang et al. 2013).

Machine learning (ML) techniques have become indispensable in the validation of such findings and modelling intricate relationships in geotechnical engineering. By leveraging ML, researchers can integrate various parameters, ensuring greater prediction accuracy for complex engineering properties (Khajehzadeh et al. 2022b; Lin et al. 2021). This capability makes ML an essential tool in predicting the bearing pressure (*P*). Its proficiency in detecting patterns, outliers, and inconsistencies ensures enhanced precision and speed in analytical workflows (Dehghanbanadaki and Motamedi 2024; Keawsawasvong et al. 2023; Singh and Roy 2023). The prediction is a critical aspect of geotechnical engineering, with significant implications for safety and cost-efficiency in construction projects (Baghbani et al. 2022; Phoon and Zhang 2023; Zhang et al. 2021). However, there exists a notable gap in the application of ML methods to a dataset encompassing diverse literature on plate load test results for shallow foundations. This gap arises from the fact that most studies are constrained to localised or site-specific datasets, which limits the generalizability of the findings across diverse geotechnical conditions.

The need to apply ML techniques to a diverse literature database is driven by several key reasons. First, shallow foundation behaviour is influenced by a wide range of variables, including soil properties, foundation geometry, and loading conditions, which vary significantly across regions. By using a diverse literature database, researchers can ensure that the models account for a broader spectrum of conditions, thereby improving their robustness and reliability. Second, the use of such datasets facilitates the development of ML models that can identify complex, non-linear relationships between variables, which may not be apparent in smaller or less diverse datasets. This capability is particularly valuable in geotechnical engineering, where the interplay of multiple factors often governs foundation performance. Such databases also enable the assessment of the relative importance of input variables across different regions, providing deeper insights into the factors influencing UBC. Finally, the application of ML to a global database aligns with the growing emphasis on creating generalised, transferable models that can be applied universally rather than requiring recalibration for each specific site. This approach not only saves time and resources but also ensures that predictions are consistent and reliable across projects with varying geotechnical conditions.

Recent efforts by Shahin (2015) highlighted the successful application of AI techniques in modelling shallow foundations, with studies often relying on small-scale laboratory datasets, site-specific observations, or numerically simulated datasets. (Sultana et al. 2022) have attempted to expand the dataset size and diversity by incorporating results from multiple literature sources. The study demonstrates that larger, more diverse datasets can significantly improve the performance and generalizability of predictive models. Artificial Neural Networks (ANNs) were adopted due to their strong capability in modelling nonlinear relationships in the study. Despite their success, ANNs present challenges such as determining optimal architectures and susceptibility to local minima, which may limit their ability to generalise complex geotechnical behaviours. (Yousuf et al. 2024) proposed exploring alternative AI models to address these limitations. When dealing with a particular geotechnical problem, various ML techniques exhibit different levels of predictive performance. By carefully choosing the appropriate algorithms, the accuracy of forecasts can be significantly enhanced (Zhang et al. 2022). Ensemble learning techniques emerge as a robust solution to the constraints of single-model approaches like ANN. These methods enhance predictive accuracy, stability, and reliability by aggregating predictions from multiple models. Some of the prominent ensemble techniques are Adaptive Boosting (AdaBoost), Gradient Boosting (GBoost), Random Trees and Extremely Randomised Trees. These techniques utilise methods like bagging and boosting to improve model performance. In geotechnical engineering, ensemble methods have been applied successfully to diverse problems: (Jas and Dodagoudar 2023) assessed liquefaction potential considering soil permeability, and (Chen et al. 2022) forecasted uniaxial tensile strength in expansive soils. Ensemble learning has proven its versatility in addressing various geotechnical properties, including soil strength, settlement, and ground motion amplification (Kim et al. 2021).

To further optimise these models, the Bayesian Optimisation Algorithm (BOA) has been introduced for efficient hyperparameter tuning. One of the critical challenges in machine learning models is the risk of overfitting, where the model captures noise in the training data, leading to poor generalisation on unseen data. (Kumar et al. 2023). Traditional hyperparameter tuning methods, such as Grid Search and Random Search, are computationally expensive and often inefficient in identifying the optimal hyperparameter configuration. To address this limitation, Bayesian Optimisation was employed as an advanced optimisation technique to systematically fine-tune the model's hyperparameters. This iterative approach enhances the model's generalisation ability, reducing overfitting and improving predictive performance while maintaining computational efficiency.

BOA utilises probabilistic models to identify optimal configurations without exhaustive searches (Garnett 2023). Compared with ensemble models, BOA enhances adaptability and performance, as demonstrated in hybrid models like AdaBoost-BOA and GBoost-BOA. These advanced hybrid approaches can accommodate diverse variables, including footing shapes, reinforcement materials, and varying soil properties. Consequently, they present powerful tools for optimising foundation designs, predicting settlement behaviour, and assessing soil stability across diverse geotechnical applications (Zhao et al. 2024). Combining ensemble models with BOA thus signifies a transformative approach to advancing predictive accuracy and addressing the complexities inherent in geotechnical engineering.

This study provides a comprehensive overview of two ensemble learning algorithms—Adaptive Boosting (AdaBoost) and Gradient Boosting (GBoost), which are further optimised through Bayesian Optimisation (BO). The development process of these hybrid models (AdaBoost-BOA and GBoost-BOA) is detailed, followed by an in-depth analysis of their predictive capabilities, comparison between standalone and optimised models and performance assessed using statistical metrics such as R^2, VAF, Mean Absolute Error (MAE), and Root Mean Square Error (RMSE). The training and testing of diverse dataset performances of these models are compared, and the optimal hybrid model is identified and benchmarked against an Artificial Neural Network (ANN) model. In addition, new criteria presented by Golbraikh and Tropsha 2002) were checked for the external validation of the optimum model on both training and testing datasets.

4.2. Models and structure

This section details the computational models. The two ensemble learning algorithms and their structures are depicted in Figure 4.1. A brief description of these models and the BOA is given in the following sub-sections.

4.2.1 Adaptive Boosting (AdaBoost)

AdaBoost (Adaptive Boosting) is a boosting algorithm that combines multiple weak learners, typically decision stumps (a one-level decision tree), to form a strong learner by sequentially adjusting the weights of incorrectly classified samples. The following are the steps

i.) Calculate maximum error

For input data r_i and target output q_i, find the maximum error at iteration t using the weak learner $fs_t(r_i)$ in equation 4.1.

$$E_t = max \mid q_i - fs_t(r_i) \mid (i = 1,2, \dots, m) \quad (4.1)$$

here m is the number of samples.

ii.) Estimate relative error:

Compute the relative error for each sample r_i using equation 4.2

$$\epsilon_{ti} = \frac{|q_i - fs_t(r_i)|}{E_t} \quad (4.2)$$

Calculate the overall error rate:

iii.) Determine the weighted average of the relative errors across all samples using their weights W_{ti} using equation 4.3.

$$\epsilon_t = \sum_{i=1}^{n} W_{ti}\, \epsilon_{ti} \quad (4.3)$$

iv.)Determine the weight coefficient (β) of the weak learner:

The weight coefficient β_t for the weak learner in iteration t is calculated based on the error rate ϵ_t as in equation 4.4.

$$\beta_t = \log\left(\frac{1 - \epsilon_t}{\epsilon_t}\right) \quad (4.4)$$

This value measures the importance of the weak learner based on its performance.

v.) Update sample weights for the next iteration *(t + 1)*:

For each sample r_i, update its weight for the next iteration *(t + 1)* using the equation 4.5

$$W_{t+1}(r_i) = \frac{W_t(r_i)}{N_t} {\beta_t}^{1-\epsilon_{ti}} \quad (4.5)$$

Where is N_t the normalisation factor that ensures the weights sum up to 1, and ϵ_{ti} is the error for sample z_i in iteration *t*.

vi.) Normalisation factor (N_t):

The normalisation factor N_t is calculated as in equation 4.6

$$N_t = \sum_{i=1}^{m} W_t(r_i)\, {\beta_t}^{1-\epsilon_{ti}} \quad (4.6)$$

vii.) Build the final model:

The ultimate learner $F(r)$ is constructed by combining all the weak learners $fs_t(r)$weighted by their respective coefficients β_t as in equation 4.7

$$F(z) = \sum_{t=1}^{T} \ln\left(\frac{1}{\beta_t}\right) fs_t\,(r) \quad (4.7)$$

4.2.2 Gradient Boosting (GBoost)

Gradient Boosting is an ensemble learning technique that builds models sequentially. Each new model corrects the errors made by the previous models by minimising a differentiable loss function, often using a gradient descent approach. The GBoost method has achieved many successful applications, and more modified versions have been developed (Chen and Guestrin 2016; Ke et al. 2017; Prokhorenkova et al. 2018). The following are the steps

i.) Initialise the base learner

Start with an initial prediction, which minimises the loss function. If using squared error loss, the base learner $G_0(z)$ is the average of the target values in equation 4.8

$$G_0(z) = \arg min_c \sum_{i=1}^{m} L(v_i, c) = \frac{1}{m} \sum_{i=1}^{m} v_i \quad (4.8)$$

Here,$L(z)$ is the loss function (squared error), *c* is a constant, and v_i is the target value for sample *i*.

ii.) Calculate the negative gradient:

For each sample z_i, compute the negative gradient, representing the residual error between the predicted and actual values. This guides the next tree's fit in the equation 4.9

$$r_{ti} = -\frac{\partial}{\partial G_{t-1}(z_i)} L\big(v_i, G_{t-1}(z_i)\big) \quad (4.9)$$

Here, r_{ti} is the negative gradient for sample z_i at iteration *t*.

iii.)Find the optimal value for each leaf node:
For each leaf node j in the regression tree, find the best value c_{tj} that minimises the loss in that node in Equation 4.10

$$c_{tj} = \arg min_c \sum_{z_i \epsilon R_{tj}} L(v_i, G_{t-1}(z_i) + c) \tag{4.10}$$

Where, R_{tj} is the set of samples in leaf node j.

iv.) Update the strong learner:

Update the overall model Ft(z) by adding the contribution of the new tree's prediction for each region as in Equation 4.11.

$$F_t(z) - F_{t-1}(z) + \sum_{j=1}^{j} c_{tj} * I(z \epsilon R_{tj}) \tag{4.11}$$

Where, $I(z \epsilon R_{tj})$ is an indicator function, and j is the number of leaf nodes.

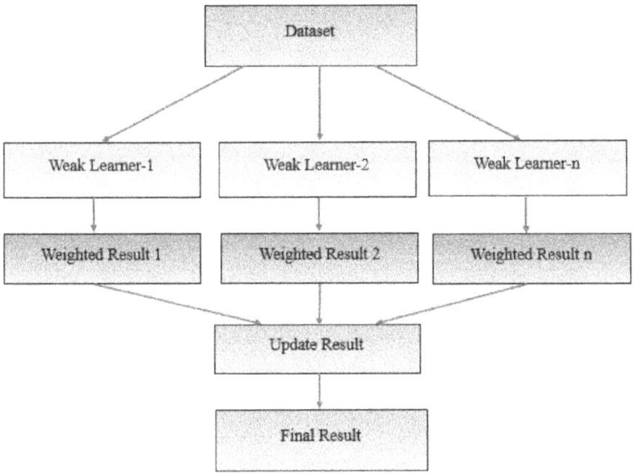

(a)

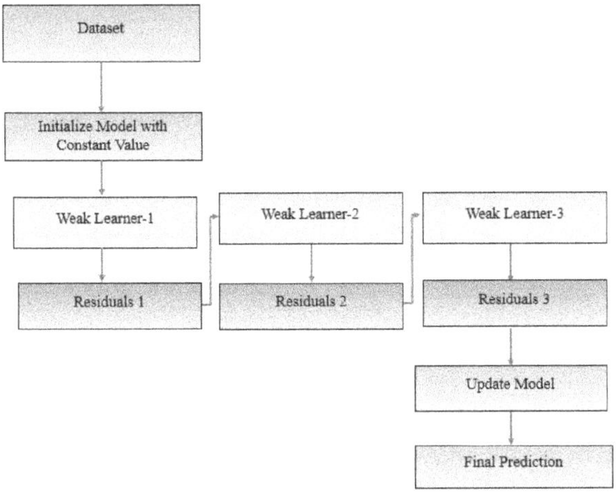

(b)

Fig 4.1. Flowchart of the ensemble learning algorithms (a) adaptive boosting method; (b) gradient boosting method

4.2.3 BOA

The Bayesian Optimisation Algorithm (BOA) is a highly efficient method for global optimisation, particularly in scenarios where prior information is available. It is composed of two key elements: (1) Gaussian Process Regression, which estimates the mean and variance of the target function across the input space, and (2) an acquisition function that guides the selection of the next point for evaluation. This combination allows BOA to balance between exploring new areas and refining promising solutions, resulting in faster convergence with fewer iterations. Due to these strengths, BOA has been applied in a range of fields where function evaluations are costly or complex. For instance, Li (2021) employed BOA to fine-tune hyperparameters for rockburst prediction. Similarly, Bo et al. (2023) utilised BOA to determine optimal parameters in an ensemble classifier designed to assess tunnel squeezing risks. Its ability to quickly find global solutions with fewer evaluations has made BOA a popular choice in optimisation problems across various domains. The algorithm is shown in Figure 4.2.

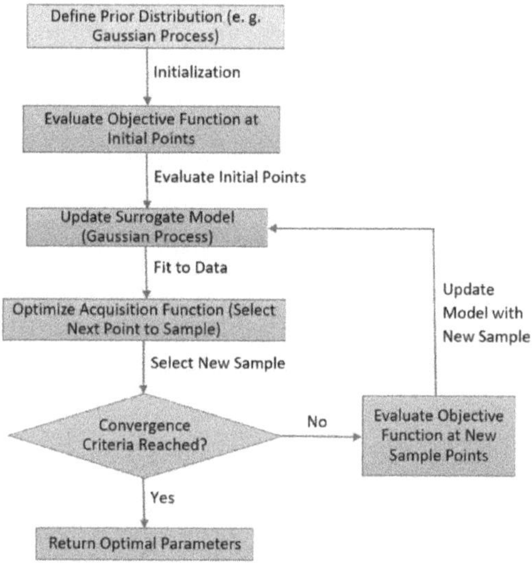

Fig 4.2. Schematic diagram of the Bayesian optimisation algorithm

4.3. Implementation process

The process begins with analysing the original dataset, followed by statistical analysis. Ensemble learning models, including AdaBoost and Gradient Boosting, are trained and optimised using Bayesian optimisation. Key predictors are used to predict UBC. A sensitivity analysis is conducted to identify the most influential parameter. The final step produces the optimal prediction model and highlights the key parameters affecting P (Bearing Pressure, the output variable). Figure 4.3 illustrates the methodology for developing an optimal prediction model for P. The implementation was carried out using Python in a Jupyter Notebook environment.

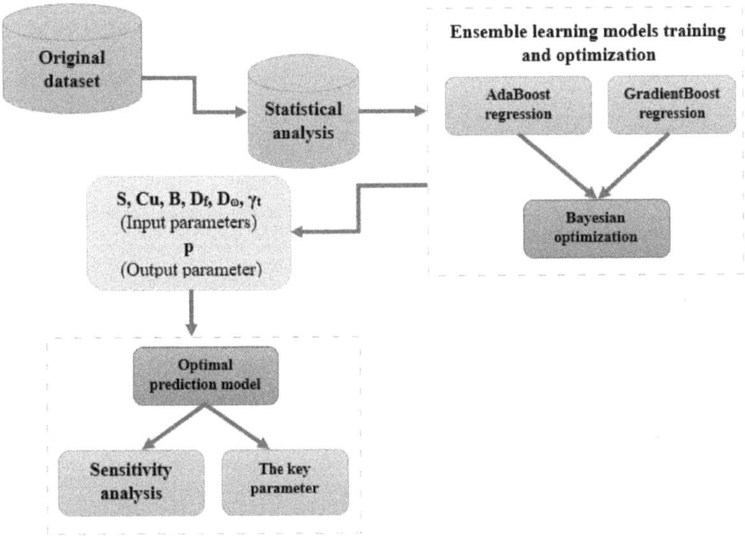

Fig 4.3. Scheme of the overall framework for bearing pressure prediction

4.3.1 Data

The dataset utilised in this study was carefully assembled from multiple reliable sources, as presented in Table 4.1. It comprises results from 58 plate load tests (PLTs) conducted on clayey soil beds across various geological settings and site conditions. Each PLT provided settlement data corresponding to 5 to 16 load levels, resulting in a total of 576 data points when all load levels were considered as individual observations. The shapes examined here were circular or square. This approach was taken to account for the variability of settlement under different load increments. The dataset includes full-scale tests performed directly in the field under natural conditions and small-scale tests conducted in controlled environments such as laboratory setups or test pits on prepared grounds. This distinction allows the dataset to encompass both real-world and simulated scenarios, ensuring broader applicability of the findings.

The input variables selected for the study include settlement (*S*), undrained cohesion (C_u), plate width or diameter (*B*), plate area (*A*), footing embedment depth (D_f), depth to groundwater table (D_w), and soil unit weight (γ_t). These parameters were chosen due to their critical role in determining the allowable bearing pressure of clayey soil. The bearing pressure (*P*) was identified as the single output variable, as it represents the ultimate measure of soil load-bearing capacity. To prepare the data for model development, training and testing data were taken in percentages of 70% and 30% of the data points, as summarised in

Table 4.2. In view of maintaining the training and testing dataset's consistency, the area (A) of footing was found to have a weaker correlation with the output (Sultana et al. 2022) in this particular dataset. Consequently, A was excluded from the analysis, resulting in the final set of input parameters (S, C_u, B, D_f, D_w, γ_t) that define the output (P). An overview of all the variables is given in Table 4.2.

Table 4.1. An overview of all the Literature from which data is taken

Source	Full scale/Small scale	Consistency	Number of Tests
Andersen and Stenhamar (1982)	Full scale	Medium	2
Bauer et al. (1976)	Full scale	Stiff to Very Stiff	2
Bergado et al. (1984)	Full scale	Medium	1
Brand et al. (1972)	Full scale	Soft	5
Dash and Bora (2013)	Small scale	Very soft	1
Desmukh and Ganpule (1994)	Full scale	Soft	2
Ghazavi and Afshar (2013)	Small scale	Soft	2
Greenwood (1975)	Full scale	Medium	1
Jardine et al. (1995)	Full scale	Soft	1
Lehane (2003)	Full scale	Soft	1
Marsland and Powell (1980)	Full scale	Very stiff	1
Newton (1975)	Full scale	Soft	3
Ornek et al. (2012)	Full scale	Stiff	7
Stuedlein and Holtz (2010)	Full scale	Stiff	3
Sultana and Dey (2019)	Small scale	Very soft	18
Tand et al. (1986)	Full scale	Medium to Stiff	8

Table 4.2. An overview of all the variables

Statistical Index	Type	P (kPa)	S (mm)	C_u (kPa)	B (m)	A (m^2)	D_f (m)	D_w (m)	γ_t (kN/m^3)
Mean	Training	162.87	22.49	37.78	0.61	0.66	0.46	1.34	17.28
Mean	Testing	163.52	22.09	39.2	0.59	0.63	0.43	1.41	17.15
Standard deviation	Training	156.2	28.12	28.01	0.57	1.55	0.69	2.17	2.62
Standard deviation	Testing	160.37	30.09	29.22	0.56	1.53	0.67	2.27	2.65
Minimum	Training	0.75	0.16	5	0.06	0.003	0	0	13
Minimum	Testing	1.48	0.17	5	0.06	0.003	0	0	13
Maximum	Training	676.35	147.83	139	3.1	9.61	2.6	10	22
Maximum	Testing	672.58	159.01	139	3.1	9.61	2.6	10	22

4.3.2 Model training

The training codes for this research are implemented using Python. All the machine learning algorithms mentioned, such as AdaBoost and GradientBoost, are executed using the Scikit-learn library. Scikit-learn, as described by Pedregosa et al. 2011), provides a robust framework for implementing these models efficiently. This library simplifies the process of training, optimising, and evaluating machine learning models.

4.4. Results and Analysis

4.1 Evaluation indices

Four key metrics—RMSE, MAE, VAF, and R^2—were used to assess model performance. Lower values of RMSE and MAE indicate a model with better predictive accuracy. RMSE emphasises larger errors, penalising significant deviations more than MAE, which gives an average error size. Both should be minimised for optimal performance.

On the other hand, R^2 measures how well the model explains the variance in the data. A higher R^2 value, ideally close to 1, shows that the model captures more variability in the observed data, indicating a stronger fit.VAF represents the percentage of the total variance in the observed data that is explained by the model's predictions. A higher VAF value indicates better predictive performance.

Thus, the best models have lower RMSE and MAE, alongside a higher R^2 and VFA,

$$RMSE = \sqrt[2]{\frac{1}{p}\sum_{a=1}^{p}(z_a - z'_a)^2} \tag{4.12}$$

$$MAE = \frac{1}{p}\sum_{a=1}^{p}|z_a - z'_a| \tag{4.13}$$

$$R^2 = 1 - \frac{\sum_{a=1}^{p}(z_a - z'_a)^2}{\sum_{a=1}^{p}(z_a - Z)^2} \tag{4.14}$$

$$VAF = \left(1 - \frac{Var(z_a - z'_a)}{Var(z_a)}\right) \times 100 \tag{4.15}$$

where z z'_a, and Z represent the target value, the prediction result, and the average of all the target values, respectively.

4.4.2 Optimisation of the hybrid models

In optimising the hyperparameters of ensemble models, the Bayesian Optimisation Algorithm is applied over 500 iterations to find the best results. AdaBoost's stability depends largely on optimising the learning rate, the depth of the decision tree, and selecting an appropriate loss function. Meanwhile, GBoost involves tuning the number of estimators, learning rate, decision tree depth, and maximum features. Each of these models relies on the correct combination of parameters to achieve accurate and stable results, and BOA is used to handle this complex optimisation task efficiently. The optimised hyperparameters are listed in Table 4.3 for both models.

Table 4.3. Hyperparameters are optimised values for different machine learning models.

Models	Hyperparameters	Optimal value	Search space
ABR	Learning rate	0.43	0.01-1
	Maximum number of estimators	89	50-200
GBR	Maximum number of estimators	87	50-200
	Maximum depth of the tree	5	1-20
	Learning rate	0.27	0.01-1
	Maximum features	13	2-20

In Figure 4.4, the comparative analysis of the AdaBoost-BOA and GBoost-BOA models highlights significant differences in their predictive performance, both on the training and testing datasets. On the training dataset, GBoost-BOA achieved an exceptionally low Mean Absolute Error (MAE) of 1.71 and Root Mean Squared Error (RMSE) of 1.18, whereas AdaBoost-BOA recorded substantially higher MAE and RMSE values of 35.07 and 43.27, respectively. These metrics indicate that GBoost-BOA provides a much closer approximation to the actual values and exhibits fewer large deviations, as reflected in its lower error values. Furthermore, the coefficient of determination (R^2) for GBoost-BOA was 99%, signifying a perfect fit to the training data, while AdaBoost-BOA demonstrated a comparatively lower R^2 of 92%, indicating that it explains less of the variance in the target variable. The Variance Accounted For (VAF) values further corroborate these observations, with GBoost-BOA achieving 99.65% compared to AdaBoost-BOA's 92.78%, emphasising the superior explanatory power of GBoost-BOA for the training data.

On the testing dataset, the performance gap between the two models persists, demonstrating GBoost-BOA's superior generalisation ability. GBoost-BOA achieved an MAE of 13.15 and an RMSE of 20.33, which are significantly lower than the corresponding values of 34.87 and 45.53 for AdaBoost-BOA. These results indicate that GBoost-BOA produces more accurate predictions and is less prone to large errors when applied to unseen data. Additionally, the R^2 value for GBoost-BOA was 98%, reflecting its ability to account for nearly all the variance in the testing dataset, whereas AdaBoost-BOA recorded a lower R^2 of 91%. The VAF values on the testing data align with this trend, with GBoost-BOA achieving 98.39 compared to AdaBoost-BOA's 92.48. These findings underscore GBoost-BOA's ability to maintain high predictive accuracy and consistency across both training and testing datasets, making it the superior model in this study. The results demonstrate that GBoost-BOA not only fits the training data with high precision but also generalises effectively to new, unseen data, which is a critical requirement for robust machine learning models in practical applications.

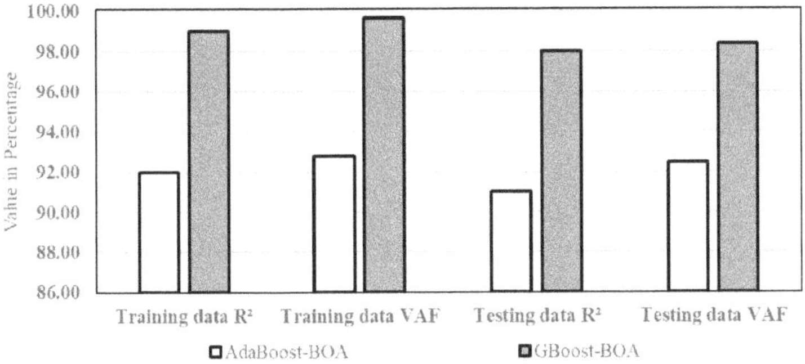

(a)

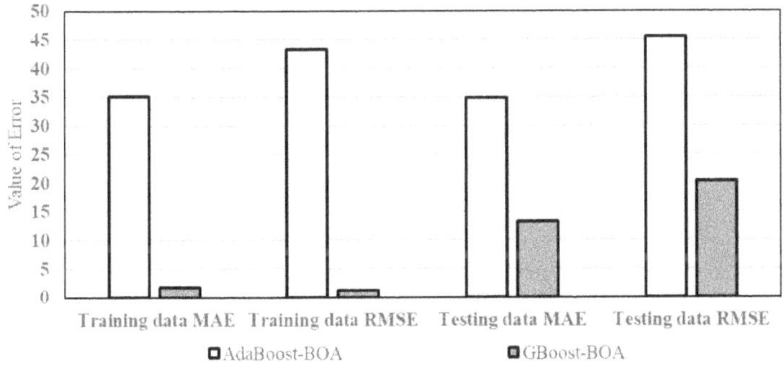

(b)

Fig 4.4 Evaluation results of the two optimised models on (a)Accuracy parameters, (b) Error parameters.

4.4.3 Comparison between hybrid models and standalone models

Figure 4.5 offers a quantitative comparison of the models using performance metrics such as R^2, VAF, RMSE, and MAE. Across all hybrid models, R^2 and VAF values are notably higher than those of the standalone models, indicating better fit and predictive strength. Conversely, both RMSE and MAE values are consistently lower for hybrid models, demonstrating reduced prediction error. Overall, based on these evaluation metrics, the hybrid models outperform their standalone counterparts, providing more reliable and precise predictions.

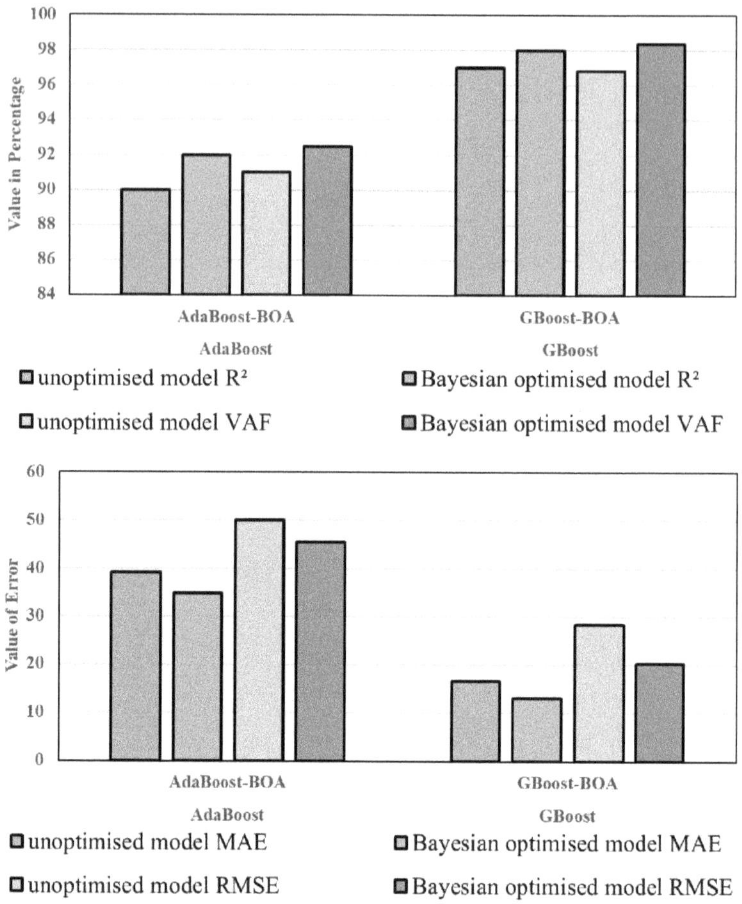

Fig 4.5 Evaluation comparison of the standalone and hybrid models on R^2, VAF, MAE, and RMSE, respectively.

Figure 4.6 presents a comparative analysis of the predictive performance of both standalone ensemble learning models and their hybrid counterparts on the test dataset. In this Figure, the "X" symbol marks the predictions generated by standalone models, while the diamonds correspond to the predictions from hybrid models. The black line serves as a reference for the ideal prediction. The distribution of the marks reveals that the predictions from the hybrid models (diamonds) are closer to the reference line compared to those from the standalone models (X marks), illustrating the superior accuracy and robustness of the hybrid models. This highlights the effectiveness of integrating optimisation algorithms to enhance predictive capabilities.

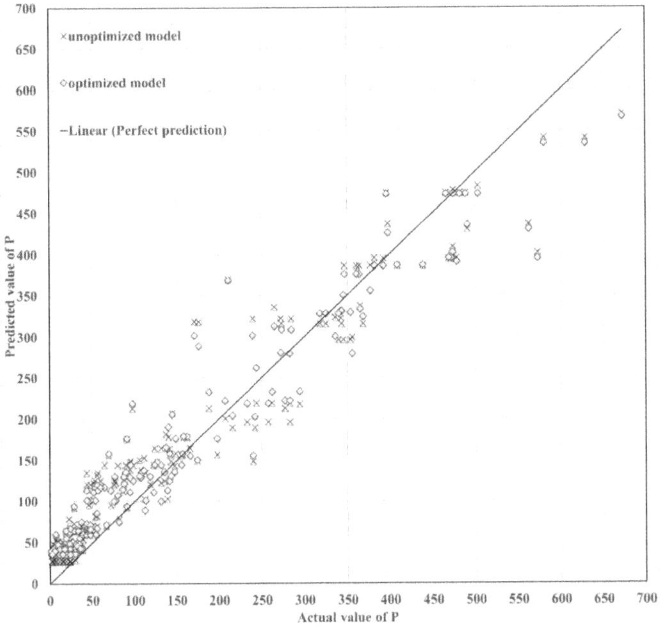

(a)

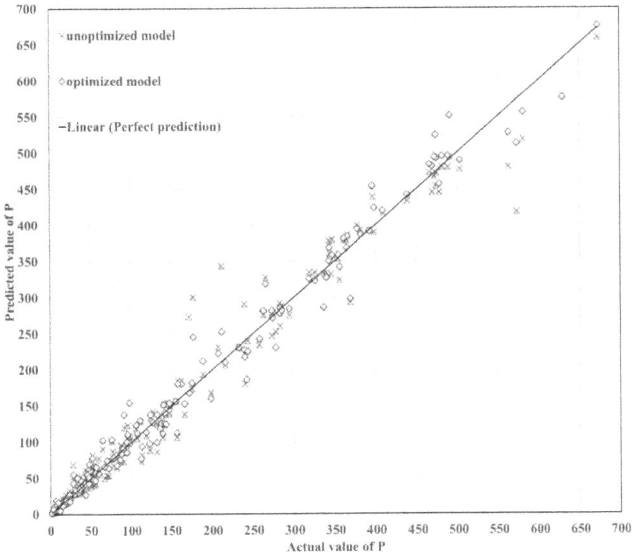

(b)

Fig 4.6 Comparison results of standalone and hybrid model prediction performance on Testing dataset (a) AdaBoost/AdaBoost-BOA(b) GBoost/GBoost-BOA

4.5 Discussion

To mitigate the issue of overfitting in the proposed model, several robust strategies aimed at ensuring generalisation and improving model reliability were applied. One of the primary measures involved utilising a large and diverse dataset for training, which encompassed a wide range of scenarios and sample variations. This approach facilitated the model's ability to capture the underlying patterns and complexities within the data, thereby reducing the risk of memorising training-specific features. Additionally, hyperparameter optimisation was conducted to fine-tune the model's parameters systematically, ensuring optimal performance without overfitting. The results demonstrated comparable performance metrics between the training and testing phases, with no significant discrepancies observed. This consistency in performance across datasets highlights the model's robustness and ability to generalise effectively. Collectively, these strategies underscore the commitment to addressing overfitting and contribute to the credibility and applicability of the proposed model hyperparameters in diverse settings. (Kumar et al. 2023)

External validation criteria are crucial for evaluating the predictive accuracy and generalisation capability of machine learning models. These criteria, as outlined by Golbraikh and Tropsha (2002), provide standardised benchmarks to assess whether a model can reliably replicate experimental outcomes and meet predefined objectives. In this study, external validation metrics were computed to evaluate the GBoost-BOA model's performance in predicting the UBC of spread footing. The results, presented in Table 4.4 for both the training and testing phases, demonstrate that all metrics fall within acceptable ranges, confirming the robustness and reliability of the models. Specifically, the experimental and predicted UBC were compared using key parameters, including k and k', the slopes of actual versus predicted and predicted versus actual regressions, respectively, constrained to pass through the origin. Additionally, coefficients of determination (R^2, R_o^2, $R_o'^2$) and the modified coefficient of determination (R_m) were calculated to measure the models' predictive power. The parameters m and n further quantified the deviation of origin-based coefficients from the overall R^2. The high values of these metrics, along with slopes close to unity, indicate a strong agreement between predicted and experimental values. These results affirm that the GBoost-BOA model is not only accurate but also consistent in predicting the UBC of spread footing, thus meeting the required standards of external validation.

$$k = \frac{\sum_{i=0}^{n}(B_i * q_i)}{\sum_{i=0}^{n}(q_i^2)} \tag{4.16}$$

$$k' = \frac{\sum_{i=0}^{n}(B_i * q_i)}{\sum_{i=0}^{n}(B_i^2)} \tag{4.17}$$

$$R_o^2 = 1 - \frac{\sum_{i=1}^{n} q_i^2 \, (1-k)^2}{\sum_{i=1}^{n}(q_i - \bar{q})} \tag{4.18}$$

$$R_o'^2 = 1 - \frac{\sum_{i=1}^{n} B_i^2 \, (1-k')^2}{\sum_{i=1}^{n}(B_i - \bar{B})} \tag{4.19}$$

$$R_m = R^2 * \left(1 - \sqrt{R^2 - R_o^2}\right) \tag{4.20}$$

$$m = \frac{R^2 - R_o^2}{R^2} \quad (4.21)$$

$$n = \frac{R^2 - R_o'^2}{R^2} \quad (4.22)$$

Table 4.4. Statistical parameters of the GBoost-BOA models for the external validation.

Parameters	Criteria	Training data	Testing data
k	$0.85 < k < 1.15$	0.9997	1.0004
k'	$0.85 < k < 1.15$	1.0002	0.9918
R_o^2	Close to 1	1	1
$R_o'^2$	Close to 1	1	0.999
R_m	$R_m > 0.5$	0.9890	0.8587
|m|	$\|m\| < 0.1$	-0.0001	-0.0164
|n|	$\|n\| < 0.1$	-0.0001	-0.0163

4.5.1 Comparison of results between the GBoost-BOA model and ANN from previous literature

Table 4.5 demonstrates that the GBoost-BOA model significantly outperforms the previously employed Artificial Neural Network (ANN) on the same training and testing data by Sultana et al. (2022). The metrics used by them to evaluate the performance of the ANN include the accuracy metrics, correlation coefficient (R) and VAF (Variance Accounted For). Additionally, they utilised the following error metrics: RMSE and MAPE (Mean Absolute Percentage Error). The results highlight the superior performance of GBoost-BOA compared to ANN in both training and testing phases. These results confirm that GBoost-BOA not only provides higher accuracy but also minimises prediction errors more effectively, making it a more robust and reliable model for soil bearing capacity prediction.

Table 4.5 Accuracy Matrix of GBoost-BOA and ANN models in training and testing phases.

Model	R (%)	VAF (%)	RMSE (*kPa*)	MAPE
ANN (training)	98.99	97.99	22.10	0.15
GBoost-BOA (training)	99.82	99.65	1.18	0.017
ANN (testing)	97.45	94.88	36.19	0.21
GBoost-BOA (testing)	99.20	98.39	20.33	0.15

4.5.2 Sensitivity

"To improve the interpretability and transparency of machine learning models, several explainable AI (XAI) techniques have been introduced in recent years (Arrieta et al., 2020; Linardatos et al., 2020). These methods are generally classified as either model-agnostic or model-specific. Model-agnostic tools—such as Local Interpretable Model-agnostic Explanations (Ribeiro et al., 2016) and SHapley Additive exPlanations (SHAP; Lundberg and Lee, 2017) can be applied to any type of ML model. They work by examining how small perturbations to the input data affect the model's output, enabling them to approximate the contribution of each feature without accessing the model's internal structure. In contrast, model-specific approaches, including Layer-wise Relevance Propagation (LRP; Bach et al., 2015), are designed for particular model architectures and exploit their internal computation pathways. Collectively, these XAI methods help clarify how models reach their predictions, highlight key influencing variables, and support the integration of domain expertise to refine model performance." Explainable AI offers tools that help clarify how complex machine learning models behave when applied to geotechnical data, where transparency is essential for engineering judgment. Local methods such as LIME focus on explaining individual predictions by approximating the model's behaviour around a specific data point, which is helpful when site conditions vary. Global methods like permutation feature importance (PFI) evaluate how much each variable contributes to overall model performance, while partial dependence plots (PDP) show the average effect of a variable on the predicted response. Together, these techniques provide complementary views of model sensitivity and help link data-driven patterns back to geotechnical understanding in a clear and defensible way. SHAP uses an additive explanation model in which feature contributions are computed with respect to a chosen background distribution; both the choice of background and the treatment of correlated features can strongly influence the resulting attributions. It provides additive feature contributions that sum to the model output, along with interaction values that decompose pairwise effects. Although SHAP is widely used for its consistent and locally accurate attributions, several limitations have been reported, including sensitivity to feature correlation, limited causal interpretation when features are dependent, and concerns about applying Shapley axioms in heterogeneous domains such as biomedicine. Its main strengths lie in offering both global and local explanations and in identifying positive or negative effects of each input feature. SHAP treats prediction as a cooperative game and assigns each feature a Shapley value, allowing clear assessment of how individual features influence model behaviour (Lundberg and Lee, 2017).

LIME, introduced by (Ribeiro et al. 2016), explains individual predictions by fitting a simple, interpretable model around a specific instance using locally simulated samples. This local approximation offers an intuitive view of how the black-box model behaves in the neighbourhood of that instance. However, its explanations depend heavily on the sampling strategy and kernel width, which can lead to instability across runs and may fail to reflect global structure, especially when features are correlated or the model is highly non-linear (Watson 2022). A formal analysis by Garreau and Luxburg (2020) further highlights both the strengths and limitations of LIME. It remains widely used because it works across tabular, image, and text data and can explain predictions from complex models, including deep networks and LLMs. Partial Dependence Plots (PDPs), introduced by Friedman (Friedman 2001), show the average effect of one or more features on model predictions by marginalising over the remaining features. Although PDPs can miss important interactions and may oversimplify the underlying relationships, they still provide useful insights when interactions are limited or low-order, and they support both single- and multi-class problems. Permutation Feature Importance (PFI) provides a global measure of feature relevance by comparing model performance before and after randomly shuffling a single feature (Peng Y.and Unluer C. 2021). The drop or change in model score reflects how strongly the permuted feature influences predictions. Each feature is permuted multiple times to produce a stable estimate of its importance. The SHAP (SHapley Additive exPlanations) value plot in Figure 8 visualises the sensitivity of different input variables in predicting bearing pressure (P) for the GBoost-BOA model. The variables ranked by their significance are: "C_u", "S", "B", "D_w", "γ_t" and "D_f". The variable "C_u" exhibits the highest SHAP value, suggesting it has the most significant impact on the model's predictions. "S" follows closely, demonstrating substantial influence, while "B" and "D_w" show moderate contributions. The lower SHAP values for "γ_t" and "D_f" suggest these variables have a comparatively lesser impact. These insights enable the prioritisation of critical variables in designing predictive frameworks and optimising geotechnical applications. This analysis emphasises the importance of feature interpretability in advanced machine learning models for practical engineering problems.

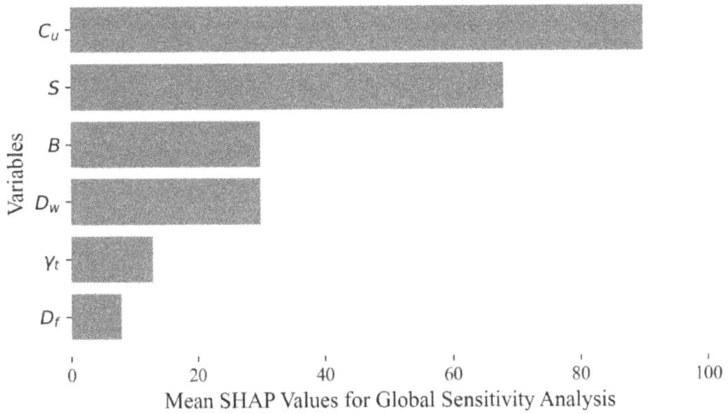

Fig 4.8 Sensitivity analysis results

4.6 Conclusion

This study provides a comprehensive evaluation of two ensemble learning algorithms, Adaptive Boosting (AdaBoost) and Gradient Boosting (GBoost), optimised through Bayesian Optimisation (BO) for predicting the bearing pressure (P) of spread foundations on clayey soil. The hybrid models, AdaBoost-BOA and GBoost-BOA, were rigorously assessed using a diverse dataset derived from 58 plate load tests (PLTs) conducted across diverse datasets. Performance metrics, including R^2, variance accounted for (VAF), root mean square error (RMSE), and mean absolute error (MAE), confirmed the superiority of GBoost-BOA over AdaBoost-BOA and conventional models. A key contribution of this research lies in the hyperparameter selection process, where BO systematically optimised the model architecture based on a diverse, multi-environmental dataset. The optimized hyperparameters for GBoost-BOA- estimators, a maximum tree depth of 89, a learning rate of 0.27 max depth of tree 5, and a maximum of 13 features—The robustness of these hyperparameters ensures that GBoost-BOA maintains high predictive performance not only on the current dataset but also when applied to other foundation engineering problems with similar geotechnical variability. This transferability is particularly critical in geotechnical engineering, where data heterogeneity often leads to overfitting in conventional models.

To mitigate overfitting, several robust strategies were implemented, including the use of a large and diverse dataset encompassing a wide range of soil conditions and foundation characteristics. This approach allowed the model to capture underlying patterns rather than memorise dataset-specific features. Additionally, hyperparameter tuning through BO ensured a well-balanced model, preventing excessive complexity while maintaining optimal accuracy. The consistency in performance across training and testing datasets, with no significant discrepancies in evaluation metrics, further validated the model's generalisation capability. The external validation using Golbraikh and Tropsha's criteria confirmed the model's generalisation and reliability, while SHAP analysis identified undrained cohesion (C_u) and settlement (S) as the most influential features. This interpretability strengthens the model's practical utility, allowing engineers to focus on the most critical soil parameters for foundation design. A comparative analysis with an Artificial Neural Network (ANN) from a previous study further reinforced the advantages of GBoost-BOA. GBoost-BOA achieved significantly higher accuracy (R^2 = 98% & R=99.20 on testing), lower RMSE (20.33), and reduced MAE (13.15), demonstrating enhanced stability across varying datasets.

In conclusion, this research establishes GBoost-BOA as a highly accurate, generalizable, and interpretable predictive model for bearing pressure (P) estimation. The incorporation of Bayesian Optimisation ensures that the hyperparameters are not merely dataset-specific but widely applicable to different foundation problems, mitigating overfitting and enhancing adaptability. By systematically optimising parameters within a diverse dataset, the model demonstrates high transferability, ensuring its effectiveness in varying geotechnical conditions. These findings highlight the potential of BO-optimised ensemble learning techniques in geotechnical engineering, paving the way for future studies on soil-structure interaction and foundation.

References

Andersen, Knut H., and Per Stenhamar. "Static plate loading tests on overconsolidated clay." Journal of the Geotechnical Engineering Division 108.7 (1982): 918-934.

Bach, Sebastian, et al. "On Pixel-Wise Explanations for Non-Linear Classifier Decisions by Layer-Wise Relevance Propagation." PLOS ONE, edited by Oscar Deniz Suarez, vol. 10, no. 7, July 2015, p. e0130140. Crossref, https://doi.org/10.1371/journal.pone.0130140.

Baghbani, Abolfazl, et al. "Application of artificial intelligence in geotechnical engineering: A state-of-the-art review." Earth-Science Reviews 228 (2022): 103991.https://doi.org/10.1016/j.earscirev.2022.103991.

Barredo Arrieta, Alejandro, et al. "Explainable Artificial Intelligence (XAI): Concepts, Taxonomies, Opportunities and Challenges toward Responsible AI." Information Fusion, vol. 58, June 2020, pp. 82–115. Crossref, https://doi.org/10.1016/j.inffus.2019.12.012.

Bauer, G. E., et al. "Predicted and observed footing settlements in a fissured clay." Proceedings of the Conference on Performance of Building Structures. Vol. 302. London: Pentech Press, 1976.

Bergado, D. T., G. Rantucci, and S. Widodo. "Full scale load tests of granular piles and sand drains in the soft of Bangkok clay." Renforcement en place des sols et des roches. Colloque international. 1984.

Bo, Yin, et al. "Robust model for tunnel squeezing using Bayesian optimised classifiers with partially missing database." Underground Space 10 (2023): 91-117.https://doi.org/10.1016/j.undsp.2022.11.001.

Brand, Edward W., Chai Muktabhant, and Anan Taechathummarak. "Load tests on small foundations in soft clay." Performance of earth and earth-supported structures. ASCE, 1972.

Chen, Guanxing, et al. "Adaptive boost approach for possible leads of triple-negative breast cancer." Chemometrics and Intelligent Laboratory Systems 231 (2022): 104690.https://doi.org/10.1016/j.chemolab.2022.104690.

Chen, Tianqi, and Carlos Guestrin. "Xgboost: A scalable tree boosting system." Proceedings of the 22nd acm sigkdd international conference on knowledge discovery and data mining. 2016.https://doi.org/10.1145/2939672.2939785.

Das, Braja M., and Nagaratnam Sivakugan. Principles of foundation engineering. Cengage learning, 2018.

Dash, Sujit Kumar, and Mukul Chandra Bora. "Improved performance of soft clay foundations using stone columns and geocell-sand mattress." Geotextiles and Geomembranes 41 (2013): 26-35.https://doi.org/10.1016/j.geotexmem.2013.09.001.

De Beer, E. E. "The scale effect on the phenomenon of progressive rupture in cohesionless soils." Soil Mech & Fdn Eng Conf Proc/Canada/. 1965.

Dehghanbanadaki, Ali, and Shervin Motamedi. "Bearing capacity prediction of shallow foundation on sandy soils: a comparative study of analytical, FEM, and machine learning approaches." Multiscale and Multidisciplinary Modelling, Experiments and Design 7.2 (2024): 1293-1310.https://doi.org/10.1007/s41939-023-00280-8.

Deshmukh, A. M., and V. T. Ganpule. "Influence of flexible mat on settlements of marine clay." Vertical and Horizontal Deformations of Foundations and Embankments. ASCE, 1994.

Fabian, Pedregosa. "Scikit-learn: Machine learning in Python." Journal of machine learning research 12 (2011): 2825.

Friedman, Jerome H. "Greedy Function Approximation: A Gradient Boosting Machine." The Annals of Statistics, vol. 29, no. 5, Oct. 2001. Crossref, https://doi.org/10.1214/aos/1013203451.

Fu, Yanwei, et al. "Unified Lexical Representation for Interpretable Visual-Language Alignment." Advances in Neural Information Processing Systems 37, 2024, pp. 1141–61. Crossref, https://doi.org/10.52202/079017-0035.

Garnett, Roman. Bayesian optimisation. Cambridge University Press, 2023.

Garreau, Damien. "Theoretical Analysis of LIME." Explainable Deep Learning AI, 2023, pp. 293–316. Crossref, https://doi.org/10.1016/b978-0-32-396098-4.00020-x.

Ghazavi, Mahmoud, and Javad Nazari Afshar. "Bearing capacity of geosynthetic encased stone columns." Geotextiles and Geomembranes 38 (2013): 26-36..https://doi.org/10.1016/j.geotexmem.2013.04.003.

Golbraikh, Alexander, and Alexander Tropsha. "Beware of q2!." Journal of molecular graphics and modelling 20.4 (2002): 269-276.https://doi.org/10.1016/S1093-3263(01)00123-1.

Greenwood, D. A. "Vibroflotation: rationale for design and practice." Methods of treatment of unstable ground (1975): 189-209.

Jardine, R. J., et al. "Vertical loading experiments on rigid pad foundations at Bothkennar." Géotechnique 45.4 (1995): 573-597.

Jas, Kaushik, and G. R. Dodagoudar. "Liquefaction potential assessment of soils using machine learning techniques: a state-of-the-art review from 1994–2021." International Journal of Geomechanics 23.7 (2023): 03123002.https://doi.org/10.1061/IJGNAI.GMENG-7788.

Ke, Guolin, et al. "Lightgbm: A highly efficient gradient boosting decision tree." Advances in neural information processing systems 30 (2017).

Keawsawasvong, Suraparb, et al. "Soft computing-based models for estimating the ultimate bearing capacity of an annular footing on Hoek–Brown material." Arabian Journal for Science and Engineering 49.4 (2024): 5989-6006.https://doi.org/10.1007/s13369-023-08588-w.

Khajehzadeh, Mohammad, and Suraparb Keawsawasvong. "Artificial Intelligence for Bearing Capacity Evaluation of Shallow Foundation: an Overview." Geotechnical and Geological Engineering 42.7 (2024): 5401-5424.https://doi.org/10.1007/s10706-024-02863-9.

Khajehzadeh, Mohammad, et al. "An effective artificial intelligence approach for slope stability evaluation." Ieee Access 10 (2022): 5660-5671.https://doi.org/10.1109/ACCESS.2022.3141432.

Khajehzadeh, Mohammad, Suraparb Keawsawasvong, and Moncef L. Nehdi. "Effective hybrid soft computing approach for optimum design of shallow foundations." Sustainability 14.3 (2022): 1847.https://doi.org/10.3390/su14031847.

Kim, Jongchan, Jongmuk Won, and Junghee Park. "Effects of water saturation and distribution on small-strain stiffness." Journal of Applied Geophysics 186 (2021): 104278.https://doi.org/10.1016/j.jappgeo.2021.104278.

Kumar, Manish, et al. "Soft computing-based prediction models for compressive strength of concrete." Case Studies in Construction Materials 19 (2023): e02321.https://doi.org/10.1016/j.cscm.2023.e02321.

Lehane, Barry M. "Vertically loaded shallow foundation on soft clayey silt." Proceedings of the Institution of Civil Engineers-Geotechnical Engineering 156.1 (2003): 17-26.

Li, Enming, et al. "Prediction of blasting mean fragment size using support vector regression combined with five optimization algorithms." Journal of Rock Mechanics and Geotechnical Engineering 13.6 (2021): 1380-1397.https://doi.org/10.1016/j.jrmge.2021.07.013.

Lin, Shan, et al. "Evaluation and prediction of slope stability using machine learning approaches." Frontiers of Structural and Civil Engineering 15.4 (2021): 821-833.https://doi.org/10.1007/s11709-021-0742-8.

Linardatos, Pantelis, et al. "Explainable AI: A Review of Machine Learning Interpretability Methods." Entropy, vol. 23, no. 1, Dec. 2020, p. 18. Crossref, https://doi.org/10.3390/e23010018.

Lundberg, Scott M., and Su-In Lee. "A unified approach to interpreting model predictions." *Advances in neural information processing systems* 30 (2017).

Marsland, A., and J. J. M. Powell. "Cyclic load tests on 865 mm diameter plates in a stiff clay till." Proceedings of the international symposium on soils under cyclic and transient loading, Swansea. 1980.

Newton, Vernon C. "Ultimate bearing capacity of shallow footings on plastic silt." (1974).

Omar, Maher, et al. "Developing artificial neural network models to predict allowable bearing capacity and elastic settlement of shallow foundation in Sharjah, United Arab Emirates." Arabian Journal of Geosciences 11 (2018): 1-11.https://doi.org/10.1007/s12517-018-3828-4.

Ornek, Murat, et al. "Prediction of bearing capacity of circular footings on soft clay stabilized with granular soil." Soils and Foundations 52.1 (2012): 69-80.https://doi.org/10.1016/j.sandf.2012.01.002.

Padmini, D., K. Ilamparuthi, and K. P. Sudheer. "Ultimate bearing capacity prediction of shallow foundations on cohesionless soils using neurofuzzy models." Computers and Geotechnics 35.1 (2008): 33-46.https://doi.org/10.1016/j.compgeo.2007.03.001.

Peng, Yiming, and Cise Unluer. "Analyzing the Mechanical Performance of Fly Ash-Based Geopolymer Concrete with Different Machine Learning Techniques." Construction and Building Materials, vol. 316, Jan. 2022, p. 125785. Crossref, https://doi.org/10.1016/j.conbuildmat.2021.125785.

Phoon, K.K. and Zhang, W., 2023. Future of machine learning in geotechnics. Georisk: Assessment and Management of Risk for Engineered Systems and Geohazards, 17(1), pp.7-22. https://doi.org/10.1080/17499518.2022.2087884.

Prokhorenkova, Liudmila, et al. "CatBoost: unbiased boosting with categorical features." Advances in neural information processing systems 31 (2018).

Ribeiro, Marco, et al. "'Why Should I Trust You?': Explaining the Predictions of Any Classifier." Proceedings of the 2016 Conference of the North American Chapter of the Association for Computational Linguistics: Demonstrations, 2016, pp. 97–101. Crossref, https://doi.org/10.18653/v1/n16-3020.

Shahin, Mohamed A. "A review of artificial intelligence applications in shallow foundations." International Journal of Geotechnical Engineering 9.1 (2015): 49-60.https://doi.org/10.1179/1939787914Y.0000000058.

Singh, Surya Pratap, and Amrit Kumar Roy. "Machine learning techniques to predict the dimensionless bearing capacity of circular footing on layered sand under inclined loads." Multiscale and Multidisciplinary Modelling, Experiments and Design 6.4 (2023): 579-590.https://doi.org/10.1007/s41939-023-00176-7.

Stuedlein, Armin W., and Robert D. Holtz. "Undrained displacement behaviour of spread footings in clay." Art of Foundation Engineering Practice. 2010. 653-669.https://doi.org/10.1061/41093(372)34.

Sultana, Parbin, and Ashim Kanti Dey. "Estimation of ultimate bearing capacity of footings on soft clay from plate load test data considering variability." Indian Geotechnical Journal 49.2 (2019): 170-183.https://doi.org/10.1007/s40098-018-0311-9.

Sultana, Parbin, Ashim Kanti Dey, and Dhawal Kumar. "Empirical approach for prediction of bearing pressure of spread footings on clayey soil using artificial intelligence (AI) techniques." Results in Engineering 15 (2022): 100489.https://doi.org/10.1016/j.rineng.2022.100489.

Tand, Kenneth E., Erik G. Funegard, and Jean-Louis Briaud. "Bearing capacity of footings on clay CPT method." Use of In Situ Tests in Geotechnical Engineering. ASCE, 1986.

Wang, Yinghe, Xinyi Zhao, and Baotian Wang. "LS-SVM and Monte Carlo methods based reliability analysis for settlement of soft clayey foundation." Journal of Rock Mechanics and Geotechnical Engineering 5.4 (2013): 312-317.https://doi.org/10.1016/j.jrmge.2012.06.003.

Watson, David S. "Conceptual Challenges for Interpretable Machine Learning." Synthese, vol. 200, no. 2, Mar. 2022. Crossref, https://doi.org/10.1007/s11229-022-03485-5.

Yamaguchi, M. "On the scale effect of footings in dense sand." Proc. 9th. Int. Conf. on SMFE. 1977.

Yousuf, Syed Md, et al. "A comparison of experimental and computational methods for circular footing response on lime-treated geotextile reinforced silty sand." Modeling Earth Systems and Environment 10.1 (2024): 1281-1303.https://doi.org/10.1007/s40808-023-01823-1.

Zhang, Wengang, et al. "Application of deep learning algorithms in geotechnical engineering: a short critical review." Artificial Intelligence Review (2021): 1-41.https://doi.org/10.1007/s10462-021-09967-1.

Zhang, Yuzhen, Jingjing Liu, and Wenjuan Shen. "A review of ensemble learning algorithms used in remote sensing applications." Applied Sciences 12.17 (2022): 8654.https://doi.org/10.3390/app12178654.

Zhao, Junjie, et al. "Performance evaluation of rock fragmentation prediction based on RF-BOA, AdaBoost-BOA, GBoost-BOA, and ERT-BOA hybrid models." Deep Underground Science and Engineering 4.1 (2025): 3-17.**https://doi.org/10.1002/dug2.12089**.

CHAPTER

5

Hybrid Bayesian Network Approach for Soil Groutability

5.1 Introduction

Permeation grouting is a commonly used method for enhancing ground conditions (Huang et al. 2007). Inflows of groundwater during underground construction can lead to critical issues, often resulting in construction delays, structural damage, and surface damage (Zebovitz et al 1989). To address this, microfine cement grouts have become increasingly popular for permeation grouting due to their improved groutability and their ability to fill small cracks and penetrate fine, low-permeability soils without causing groundwater pollution. One significant difficulty in utilising microfine cement grouts is predicting the ease with which the grout can be injected and distributed in the intended geological material, as the process of grouting involves the intricate and time-sensitive movement of cement particles through the soil structure (Tekin and Akbas 2011).

The literature describes multiple formula-based approaches to predict groutability. Early investigations by Burwell (1958), Mitchell (1981) introduced a straightforward and practical equation to define groutability as follows:

$$N = D_{(15)\ base\ soil}/d_{(85)\ cement\ grout} \tag{5.1}$$

Where $D_{(15)\text{ base soil}}$ is the sieve size that 15% of the soil passes through, and $d_{(85)\text{cement grout}}$ is the sieve size that 85% of the grout passes through. The successful injection of grout into the soil is feasible when N exceeds 25, while inadequate grout injection occurs when N falls below 11. For N values ranging between 25 and 11, conducting in situ tests is recommended to ascertain the feasibility of grouting.

Burwell (1958) proposed an additional equation (Eq. 5.2) to further evaluate the feasibility of grouting, even when N exceeds 25.

$$N = D_{(10)\ base\ soil}/d_{(95)\ cement\ grout} \tag{5.2}$$

Where $D_{(10)\ \text{base soil}}$ is the sieve size that 10% of the soil passes through, and $d_{(95)\text{cement grout}}$ is the sieve size that 95% of the grout passes through. Here, *N* values over 11 indicate grouting is possible, while under 5 is not, and 5 to 11 requires in situ tests.

Incecik and Ceren (1995) introduced a new equation

$$N = D_{(10)\ base\ soil}/d_{(90)\ cement\ grout} \tag{5.3}$$

Where $D_{(10)\ \text{base soil}}$ is the sieve size that 10% of the soil passes through, and $d_{(90)\text{cement grout}}$ is the sieve size that 90% of the grout passes through. If the value of *N* from Equation 5.3 exceeds 10, it implies that grouting is viable in the soil.

DeBeer (1949) proposed an equation (Equation 5.4) to predict the groutability of soil based on its hydraulic conductivity (*k*) and temperature (*t*). The equation is as follows:

$$k = 116 \times (0.7 + 0.034t) \times D_{10}^2 \tag{5.4}$$

In this equation, *k* represents the hydraulic conductivity in centimetres per second (*cm/s*), *t* denotes the temperature in degrees Celsius (°*C*), and D_{10} signifies the grain size corresponding to 10% finer by weight in centimetres (*cm*). Landry et al (2000) proposed that the obtained value of *k* from Equation (5.4) serves as an indicator of soil groutability. If *k* exceeds 1×10^{-1} cm/s, cement grouting is effective. For 1×10^{-1} *cm/s* < *k* < 5×10^{-3} *cm/s*, finer-grained cement should be used. For 5×10^{-3} *cm/s* < *k* < 1×10^{-4} *cm/s*, special chemical solutions should be used. This method quantitatively assesses soil groutability based on permeability and temperature.

Subsequent studies, including those by Akbulut and Saglamer (2002), Tekin (2004), and Kim et al (2009), have underscored the importance of incorporating additional parameters to enhance the reliability of groutability estimation.

$$N = ((D^2{}_{(10)base\ soil}))/((d^2{}_{90\ cement\ grout}) + k_1 \times ((\tfrac{\text{w}}{\text{c}})/\text{FC}) + k_2 \times (P/D_r) \tag{5.5}$$

Equation (5.5) evaluates the groutability value considering the water-cement ratio (*w/c*), fine soil content (*FC*), grouting pressure (*P*), and soil relative density (D_r). Constants k_1 and k_2, set to 0.5 and 0.01 in 1/*kPa*, respectively, adjust *N* based on tests. *N* values over 28 suggest cement grouting is effective, while lower values recommend alternatives like chemical grouting. Equation 5.5 applies best when *FC* < 6%, *w/c* between 0.8 and 2, and grouting pressure ranges from 50 to 200 *kPa*.

Predicting the groutability of a given soil-grout system is a complex task, as it depends on various site-specific factors such as soil and grout grain size distribution, grout suspension concentration and viscosity, soil pore size, hydraulic conductivity, and injection pressure (Tekin and Akbas 2011). Traditional empirical approaches often fail to provide reliable results, and performing laboratory tests or in-situ testing is costly and time-consuming (Nowroozi et al. 2021), especially for the use of nanometer-scale grouts (Tekin and Akbas 2011). To tackle this challenge, artificial intelligence (AI) methods may offer innovative ways to approach challenges. To et al. (2024); Sang et al. (2023), enabling the discovery of novel insights from previous data and adaptively updating the prediction process as new information becomes available (Cheng et al. 2013). The groutability prediction problem can be approached as a classification task with two categories: whether the grouting in the soil is a success or failure, positioning AI-driven classifiers as a possible strategy (Cheng et al. 2013). Nonetheless, implementing artificial neural network (ANN) models presents certain challenges (Ghanizadeh et al. 2023). Determining the optimal number of hidden layers, the number of neurons in each layer, and the learning rate can be complex. Additionally, the training process, which relies on gradient descent algorithms, can be complex and prone to being trapped in local minima, potentially compromising the predictive performance. The complexity of ANN models can hinder construction engineers' ability to interpret how these models arrive at groutability predictions (Movassagh et al. 2023). Bayesian classifiers, based on the principles of Bayes' decision theory, offer a promising alternative for solving groutability prediction as a classification problem (Agrawal and Bala 2008; Torres and Calderara 2023). Prior studies have indicated that Bayesian techniques can offer prediction accuracy on par with that of Artificial Neural Network (ANN) models (Rejwanur et al. 2023; Kononenko 1993; Sebastiani 2002). One such method is the Bayesian network. Bayesian Networks (BNs) provide a valuable tool for visualising complex problems and handling nonlinear issues with uncertainty. Bayesian Networks have several advantages, including incorporating prior knowledge and multiple data sources, probabilistic updates without altering the structure, facilitating decision-making in the presence of full or partial data, and offering tools and methodologies that empower effective decision-making under various data availability scenarios (Hu et al 2019). Soil grouting involves numerous factors that are continuous variables, like the water/cement ratio of the grout, relative density of the soil, Grouting pressure, etc., while the grouting potential target is a discrete variable (1 indicating successful grouting and 0 indicating unsuccessful grouting). It is required to discretise continuous variables before constructing a Bayesian network, where critical information is lost, which results in erroneous variable interdependencies (Hanif et al. 2021). Class imbalance and sampling bias pose significant challenges, as the ratio of groutable to non-groutable samples is often skewed, yet the sensitivity of existing models to these issues limits their effectiveness. Additionally, most groutability prediction studies prioritise general performance metrics without adequately emphasising recall, which is critical for identifying groutable soil samples. Lastly, the interpretability of models, computational efficiency, and adaptability to new data are often overlooked, further highlighting the need for robust methods like hybrid BN models that address these limitations while minimising information loss from data discretisation. Several Bayesian network (BN) models have been utilised in different domains, such as assessing risk in earth dams (Morales et al. 2014), forecasting earthquake-induced liquefaction (Hu et al. 2022), risk mapping (Maldonado et al. 2016), probabilistic seismic analysis (Gehl et al. 2017), etc. So far, no hybrid Bayesian network model has been observed to be employed to forecast groutability in soils. The current research presents a novel approach by developing a hybrid Bayesian Network (BN) model that integrates both continuous and discrete variables, surpassing the limitations of traditional BN models restricted to discrete variables. Employing a rigorous three-step cross-validation process with distinct training, testing, and validation datasets, the study ensures robust model validation and mitigates overfitting risks. Utilising the *bnlearn* toolkit in R, the model adeptly handles complex datasets, leveraging the hill-climbing algorithm for structure learning. The inclusion of multiple machine learning models, such as Support Vector Machine (SVM) and Naïve Bayes, facilitates comprehensive performance comparison, with the hybrid BN model demonstrating superior predictive accuracy and generalisation capabilities. By integrating domain knowledge and historical data, the study enhances the model's probabilistic reasoning. By employing a hybrid Bayesian Network (BN) model, the research fills the gap in accurately capturing the complex interdependencies between influential factors like grouting pressure, soil particle size distribution, and water/cement ratio, etc. Traditional models often fail to handle such complexity, leading to suboptimal predictions.

This study aims to present a methodology for constructing a hybrid Bayesian network model to predict the groutability potential of coarse-grained soil using the R programming language and the *bnlearn* package. The proposed approach employs a Markov Chain Monte Carlo (MCMC) method and demonstrates the superior performance of Hybrid Bayesian Network models over Discrete Bayesian Network models and other methodologies. The Hybrid BN model integrates both continuous and discrete variables, offering a comprehensive representation of the problem domain. A 3-step cross-validation process is implemented to mitigate the risk of overfitting and ensure the model's generalizability and applicability across various datasets. The effectiveness of the hybrid Bayesian network model is evaluated by comparing its performance with discrete Bayesian network models and other approaches, such as support vector machines and Naïve Bayes. To validate the generalizability of the Hybrid Bayesian Network model, it is applied to a separate validation dataset, and the results are compared to those of the other three models. This approach provides a rigorous evaluation of the model's performance and highlights its potential applicability in predicting the groutability potential of coarse-grained soil.

5.2 The hybrid Bayesian network

A Bayesian Network (BN) is a probabilistic graphical model that provides a sophisticated framework for articulating the joint probability distribution of multiple random variables. It serves as a valuable instrument for addressing the ambiguity involved in navigating uncertainties, drawing inferences from knowledge, and solving intricate nonlinear problems. The essential elements of a BN model typically include a directed acyclic graph (DAG) paired with a set of conditional probability tables (CPTs) or conditional probability distributions (CPDs). The probability distribution embedded in the Bayesian Network framework can be mathematically expressed as shown in Equation 5.6:

$$P(B1, B2, \cdots, Bn) = \prod_{i=1}^{n} (P(Bi|\pi(Bi))) \qquad (5.6)$$

Equation 5.6 in the context of Bayesian networks is a concise way to represent a joint probability distribution. In such networks, a set of random variables $\{B1, B2, ..., Bn\}$ is represented, and the joint probability distribution over these variables is expressed as a product of conditional probabilities. The left side, $P(B1, B2, ..., Bn)$, denotes the joint probability of all variables. The right side uses the chain rule of probability to express the joint probability as the product of conditional probabilities $P(Bi|\pi(Bi))$ for each variable given its parents, $\pi(Bi)$. This approach leverages the conditional independence among variables, allowing for efficient computation and representation of the joint probability distribution.

When evidence, denoted as '*f*,' is available, we can calculate the posterior probability as shown in Equation 5.7:

$$P(B|f) = \frac{P(f,B)}{P(f)} = \frac{P(f|B)P(B)}{\sum_B P(f|B)P(B)} \quad (5.7)$$

Here, $P(B|f)$ denotes the posterior probability of variable B given evidence f, a key aspect in updating beliefs. Derived from Bayes' theorem, it expresses this posterior as the joint probability of B and f divided by the marginal probability of f. The formulation incorporates the likelihood of the evidence given B, i.e. $(P(f|B))$ and the prior probability of B, i.e. $(P(B))$, crucial for probabilistic reasoning. In continuous scenarios, the denominator involves integration rather than summation. This equation plays a fundamental role in dynamically updating probabilities, accommodating new evidence, and ensuring the probability distribution over the variable's sums to 1 across all possible values of B. In hybrid Bayesian networks, accommodating both continuous and discrete variables, this equation showcases the adaptability of Bayesian networks in handling diverse data types.

Hybrid BNs offer a versatile framework for modelling complex systems that encompass both continuous and discrete variables. In Figure 5.1, elliptical nodes represent continuous variables, and rectangular nodes denote discrete variables. The arrows illustrate causal relationships, and the conditional probability tables (CPTs) and conditional probability distributions (CPDs) capture the likelihoods given the parent variables. While hybrid Bayesian networks provide flexibility in representing a wide range of real-world phenomena, their computational complexity for learning and inference can be higher than that of discrete Bayesian networks.

The modelling process for hybrid Bayesian networks consists of two primary phases: structure learning and parameter learning. Structure learning establishes the network's topology using expert insights or algorithms designed for discrete variables. More complex parameter learning involves determining the joint probability distribution for continuous variables (Asghari et al. 2021), often using Gaussian or log-normal distributions. The chain rule for Bayesian networks allows for inferring the posterior probability of a discrete node (B1) given its parent nodes' evidence (B2 and B3 in Figure 5. 1). Despite computational difficulties, hybrid BNs are powerful for inference, prediction, and diagnosis in systems with both discrete and continuous variables (Hu et al. 2016; Glymour et al 2001).

In a hybrid BN accommodating both discrete variables (set I) and continuous variables (set Y) with a presumed linear relationship, the joint probability distributions are divided into discrete and mixed components. This separation facilitates a more manageable representation of complex relationships within the network. Focusing on a node Bi and its parents $\pi(Bi)$ with j combinations of states in a Directed Acyclic Graph (DAG), the parameter learning for the discrete part is articulated through Equation 5.8.

$$\theta_{ijk} = P(B_i = k|\pi(B_i) = j) \quad (5.8)$$

This equation denotes the conditional probability of node B_i taking the value k given the specific combination of states for its parents. This parameterisation enables the modelling of the discrete aspect of the hybrid BN, providing a foundation for understanding the probabilistic dependencies in the network. The expression encapsulates the likelihood of observing discrete variable states based on the states of their parents, offering a crucial element in the comprehensive learning and inference processes of hybrid BNs.

Here, θ_{ijk} denotes the estimated parameter, and the overall parameter count is $\sum_{i^*=1}^{n} q_i r_i$, determined by the nodes, edges, and intervals. The count of independent parameters is $Q_\theta = \sum_{i^*=1}^{n} q_i (r_i - 1)$, considering $\sum\theta_{ijk} = 1$ for a node.

It is hypothesised that the local probability model follows a Gaussian distribution, with child nodes linearly correlated to continuous parent nodes, as detailed in Figure 5.1 and Equation 5.9.

$$\theta_{\gamma|\pi I(\gamma)} = (m_{\gamma|\pi I(\gamma)}, \beta_{\gamma|\pi I(\gamma)}, \sigma^2_{\gamma|\pi I(\gamma)}) \tag{5.9}$$

Equation 5.9 encapsulates the parameters governing a linear Gaussian distribution for a continuous node γ given its discrete parent nodes represented by $\pi I(\gamma)$. Specifically, $\beta_{\gamma|\pi I(\gamma)}$ signifies the regression coefficients, $m_{\gamma|\pi I(\gamma)}$, denotes the regression intercept, and $\sigma^2_{\gamma|\pi I(\gamma)}$ represents the conditional variance. The notation $\pi I(\gamma)$ refers to a set of discrete parent nodes associated with the continuous node γ. This formulation enables the expression of the Gaussian distribution for the continuous nodes in the mixed component, outlined in Equation 5.10.

$$(Y_{\gamma|\pi I(\gamma)} \pi_y(\gamma), \theta_{\gamma|\pi I(\gamma)}) \sim N\,(m_{\gamma|\pi I(\gamma)} + \beta_{\gamma|\pi I(\gamma)} \pi_y(\gamma), \sigma^2_{\gamma|\pi I(\gamma)}) \tag{5.10}$$

In Equation 5.10, the left-hand part follows a normal distribution (N) with mean $m_{\gamma|\pi I(\gamma)} + \beta_{\gamma|\pi I(\gamma)} \pi_y(\gamma)$ and variance $\sigma^2_{\gamma|\pi I(\gamma)}$. This comprehensive representation allows for modelling the relationship between continuous and discrete variables within the hybrid Bayesian Network, accounting for the regression parameters, intercept, and conditional variance in a Gaussian framework.

Subsequently, the joint probability distribution of the mixed part can be formulated as Equation 5.11:

$$f(Y|I, \theta_i) = \prod_{y=1}^{n} (f(Y_{\gamma|\pi I(\gamma)}, \pi_y(\gamma), \theta_{\gamma|\pi I(\gamma)})) \tag{5.11}$$

Here, 'n' signifies the quantity of variables within the continuous set. As a result, the model's unidentified parameter can be denoted as $\theta = (\theta_I, \theta_Y)$. Through the amalgamation of discrete and mixed components, the comprehensive model yields the joint probability distribution. (Equation 5.12):

$$P(X|\theta) = P(I, Y|\theta) = P(I|\theta_I) P(Y|I, \theta_Y) = \prod_{\delta=1}^{l} (P(I_{\delta|\pi(\delta)}, \theta_{\delta|\pi(\delta)})) \prod_{\delta=1}^{n} (f(Y_{\gamma|\pi I(\gamma)}, \pi_y(\gamma), \theta_{\gamma|\pi I(\gamma)})) \tag{5.12}$$

In this context, 'l' represents the number of variables in the discrete set.

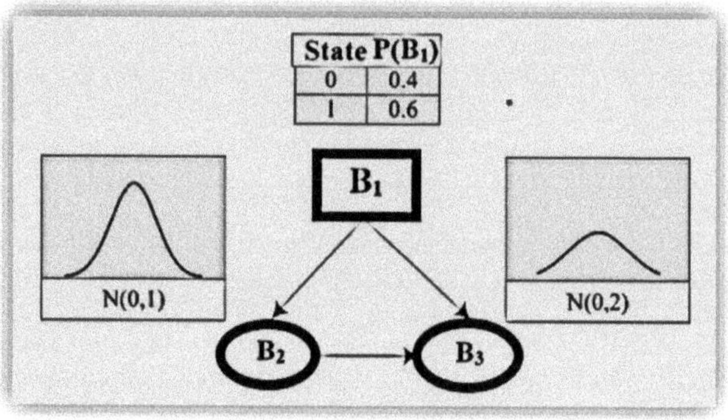

Fig.5.1. Conditional Probability Table (CPT)and two Conditional Probability Distributions (CPDs) for a hybrid Bayesian network model

5.3 Building BN models using soil grouting data

5.3.1 Data and model variables

The dataset utilised in this study has 117 data points, out of which 31 cases are from Tekin 2001. The remaining case records are from Akbulut and Saglamer (2002); Zebovitz et al. (1989); and Avci (2009). This study incorporates nine influential factors as listed in Table 5.1.

Table 5.1 Variables used in the dataset.

Factors	Description	Mean	Min	Max
X1.	The water/cement ratio of the grout	1.573	1	6
X2.	Relative density of the soil (%)	51.52	27	80
X3.	Grouting pressure (kPa)	333.5	50	690
X4.	Content of soil passing through a 0.6 mm sieve (%)	45.88	1	100
X5.	The diameter of a sieve through which 10% of the soil passes (mm)	0.3524	0.08	2.5
X6.	The diameter of a sieve through which 15% of the soil passes (mm)	0.3811	0.1	2.7
X7.	The diameter of a sieve through which 85% of the cement grout passes (mm)	0.01407	0.00075	0.031
X8.	The diameter of a sieve through which 90% of the cement grout passes (mm)	0.0157	0.0008	0.04
X9.	The diameter of a sieve through which 95% of the cement grout passes (mm)	0.02537	0.0012	0.06

In the realm of construction and geotechnical engineering, the intricate and interdependent mathematical relationships between these variables, such as grouting pressure, soil particle size distribution, and the water/cement ratio, are essential for optimising grouting processes. The fundamental role of the water/cement ratio in influencing grout properties includes strength and durability. Grouting pressure, measured in *kPa*, is intricately linked to soil density, requiring higher pressures in loosely compacted soils for effective penetration and distribution of the grout.

Understanding soil gradation is pivotal, influencing engineering properties like permeability and shear strength (Afrazi and Yazdani 2021), which, in turn, impact soil groutability. Additionally, the diameters of sieves for cement grout offer vital information on particle distribution, which is crucial for evaluating the efficacy of the grouting process in filling voids within the soil. These relationships collectively guide geotechnical engineers in optimising construction practices, ensuring structural stability, and achieving desired material properties, ultimately influencing the success of grouting processes in enhancing soil groutability.

5.3.2. Experimental design

The dataset used in this study comprises 117 data points from four different sources. This diverse sourcing ensures the inclusion of a wide variety of data points. To address concerns about the limited sample size, the hybrid Bayesian Network (BN) model was tested using three distinct training and testing data arrangements, each drawn from a pool of 100 data points. Crucially, each arrangement involved completely different test data, reducing the risk of overfitting and enhancing the model's robustness across various datasets, as noted by (Asadizadeh and Majdi 2019). Additionally, a separate validation set was used to demonstrate the model's generalizability, ensuring its reliability and potential for real-world application despite the sample size constraints. This rigorous testing process contributed to the confidence in the model's performance and its potential for real-world application.

The 117 data points were initially split into two main parts: 100 data points for further division into training and testing datasets, and the remaining 17 data points allocated specifically for validation purposes. This allocation allowed for the assessment and validation of the Hybrid BN model's generalizability. Notably, the validation dataset remained separate throughout the model training and testing phases, ensuring its independence and integrity during evaluation.

For cross-validation, the set of 100 data points designated for training and testing was divided into two groups: one group containing 83 data points for training and the other group containing 17 data points for testing. This division resulted in subsets within the database, with proportions of 71% (83 case records) for training, 14.5% (17 case records) for testing, and an equivalent 14.5% (17 case records) for validation. These proportions are typical of AI-based modelling studies (Cheng and Hoang 2014); (Cabalar et al. 2012). This segmentation in Figure 5.2 facilitated the evaluation of model performance across different subsets of data, enhancing its reliability and robustness.

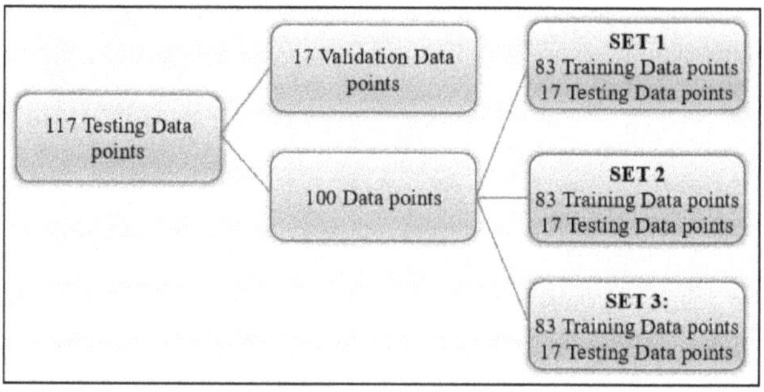

Fig 5.2 Data division for the 3-step cross-validation test

The sensitivity of the model to hyperparameters was addressed through MCMC sampling with 10,000 iterations during the inference phase. This ensured sufficient convergence and stability in the posterior distribution. No explicit hyperparameter optimisation method, such as Bayesian optimisation, was applied. Instead, we used extensive sampling to reduce sensitivity and improve performance. Performance metrics (Accuracy, Precision, Recall, AUC) were calculated and validated across both training and test datasets.

5.3.3. Structuring of the Hybrid Bayesian Network Model

The *bnlearn* toolkit is an invaluable resource in the domain of BN modelling, seamlessly integrated with the R programming language (Scutari 2009). Specifically designed for hybrid Bayesian Networks, it excels in handling datasets comprising both continuous and mixed variables, making it a versatile tool for complex modelling scenarios. This toolkit encompasses various structure learning techniques, including the hill-climbing (HC) algorithm, allowing for the effective extraction of relationships among variables. Hill-Climbing is particularly suitable for continuous and discrete hybrid Bayesian networks in geotechnical problems due to its flexibility in handling mixed data types. Additionally, its scalability and ease of implementation make it a practical choice for real-world geotechnical datasets. Hu (2022) recommends using the HC algorithm for geotechnical problems.

In the context of predicting groutability, the *bnlearn* toolkit proves instrumental. The potential for grouting (y) is aptly represented as a discrete categorical variable with binary states: "yes" indicating successful grouting and "no" denoting unsuccessful grouting. The parent nodes influencing this outcome consist of continuous variables, reflecting the multifaceted nature of the grouting process. Notably, the toolkit accommodates both linear and non-linear relationships between variables, providing a comprehensive framework for capturing the intricacies of groutability prediction.

Figure 5.3. illustrates the initial Bayesian Network structure for groutability prediction, constructed using the HC algorithm. The MCMC procedure was employed for carrying out an approximation of inference within the hybrid configuration using the MCMC-Bayesian Network framework. As shown in Figure 5.3, the provided Bayesian Network structure is a graphical model that represents probabilistic relationships among a set of variables, denoted as (y), (x1) through(x9). The presented Hybrid BN model describes a probabilistic graphical model where nodes represent random variables, and the directed edges between nodes denote conditional dependencies. The structure is indicative of an underlying causal relationship among the variables (y), (x1), (x2), (x3), (x4), (x5), (x6), (x7), (x8), and (x9). This model incorporates both discrete (e.g., (y)) and continuous (e.g., (x1), (x2)) variables. The structure has likely been determined using the HC algorithm, a local search algorithm commonly employed for learning the structure of BNs. HC iteratively refines the structure by adding or removing edges based on a scoring metric that evaluates the goodness of fit of the model to the data. In this context, the edges are added or removed to optimise the likelihood of the observed data given the model. Starting with an initial structure or an empty graph, the HC algorithm incrementally explores possible changes to the network structure, evaluating the impact of these changes on the model's fit to the data. The Edges are added or removed based on statistical tests or information criteria until an optimal structure, in terms of model fit, is reached.

The resulting BN structure indicates the conditional dependencies and influences among the variables. The variable (y) serves as a foundational node in the network, indicating a primary factor that influences the behaviour of the system. (x7) is directly conditioned on (y), suggesting that (x7) is influenced by the state or value of (y). Furthermore, (x2) is then dependent on (x7), creating a cascade of dependencies. This implies that changes in (y) may propagate through the network, impacting both (x7) and, subsequently, (x2). The conditional dependency [x1|x2: y] implies that (x1) depends on both (x2) and (y), showcasing a more intricate relationship where the state of (x1) is influenced by the joint values of (x2) and (y). The network becomes even more intricate with the introduction of variables like (x9), (x3), (x8), (x5), (x6), and (x4). For instance, (x9) is conditioned on (x1), (x7), and (y), indicating that (x9) is influenced by multiple preceding variables. (x3) depends on both (x2) and (x9), reinforcing the notion that the state of one variable depends on the values of others. The variable (x8) is particularly interesting as it is conditioned on a set of variables, including (x1), (x2), (x3), (x7), (x9), and (y), highlighting a comprehensive interdependence within the network. After this, the steps for using the MCMC to approximate the hybrid BN model inference should be performed for further results, as shown in Figure 5.4

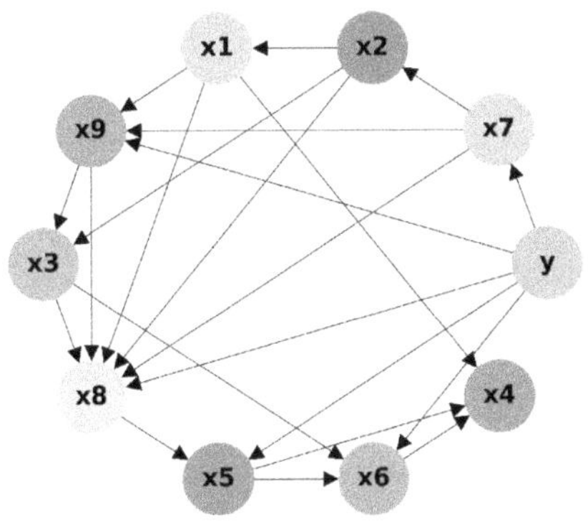

Fig 5.3. A Hybrid BN using the MCMC-BN model following the Hill Climber algorithm

Steps	Description
Step 1	Specify the network structure (as shown in Figure 5.3)
Step 2	Perform data division as mentioned in the experimental design.
1	Use HydeNetwork() function to define the network structure and training dataset.
2	Use WriteNetworkModel() function to send the network and training data to JAGS. Conduct parameter learning to: – Learn normal distribution parameters (for continuous nodes) – Learn conditional probability values (for discrete nodes)
3	Use CompileJagsModel() function to introduce the test dataset as evidence variables in JAGS.
4	Use HydePosterior() function for MCMC inference on the target node with 10,000 iterations.
5	Derive posterior distribution samples of the target node. Identify the discrete value with the highest frequency as the final predicted value.
6	Use tools like PROC, gmodels, etc., to compute classification performance metrics.
Step 3	Compute average performance measures from the MCMC-BN model using three different train-test dataset arrangements.

Fig 5.4. The steps for using the MCMC to approximate the hybrid BN model inference

5.4. More models developed based on grouting data

In this comprehensive study on predicting groutability, a variety of machine learning models were systematically applied for comparison with the Hybrid BN model. The Discrete BN model employs the HC algorithm on discretised continuous variables to create an initial model design depicted in Figure 4.5, based on the discrete database detailed in Table 5.2. The discretisation of continuous geotechnical variables was done by dividing them into five equal-width intervals or five equal parts. This decision was based on simplifying the data while preserving its relevance for analysis. Equal-width intervals were chosen to balance precision and ease of interpretation.

The BN model's parameters were then meticulously learned through the Bayesian estimation algorithm, and its performance was rigorously assessed in the same way and on the same training and testing data as the Hybrid BN model was assessed, and the average values were recorded. In parallel, the Naïve Bayes (NB) model, sharing conceptual similarities with the BN but assuming independence among dependent variables, introduced a simplification in the calculation of posterior probability. For this investigation, continuous variables in the NB model were assumed to follow a Gaussian distribution. Complementing these probabilistic models, the study also incorporated a Support Vector Machine (SVM) approach. SVM, grounded in statistical theory, is particularly adept at handling challenges posed by small sample sizes, high dimensionality, and substantial nonlinearity (Kamran et al. 2021). Both NB and SVM performance were assessed in the same way and on the same training and testing data, and average values were reported.

By integrating these diverse modelling techniques, the study sought to provide a thorough and nuanced analysis of groutability prediction, considering both discrete and continuous variables in a robust and well-validated analytical framework. The inclusion of multiple models enhances the reliability and comprehensiveness of the predictive insights derived from the investigation.

Table 5.2 Discretise values for the discrete BN models.

Factors	1	2	3	4	5
X1.	0.995-2	2-3	3-4	4-5	5-6
X2.	26.9-37.6	37.6-48.2	48.2-58.8	58.8-69.4	69.4-80.1
X3.	49.4-178	178-306	306-434	434-562	562-691
X4.	0.901-20.8	20.8-40.6	40.6-60.4	60.4-80.2	80.2-100
X5.	0.0776-0.564	0.564-1.05	1.05-1.53	1.53-2.02	2.02-2.5
X6.	0.0974-0.62	0.62-1.14	1.14-1.66	1.66-2.18	2.18-2.7
X7.	0.00072-0.006	0.0068-0.013	0.0128-0.019	0.0189-0.0249	0.0249-0.031
X8.	0.00076-0.0086	0.0086-0.016	0.0165-0.0243	0.0243-0.0322	0.0322-0.04
X9.	0.00114-0.013	0.013-0.0247	0.0247-0.0365	0.0365-0.0482	0.0482-0.0601

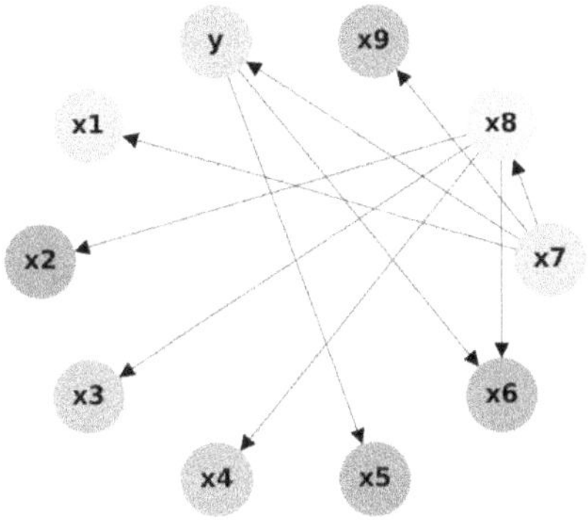

Fig 5.5. A Discrete BN model following the HC algorithm

5.5 Performance measures

In the assessment of our models' classification performance, we applied a diverse set of metrics, including accuracy, precision, recall, F1 score, and the area under the Receiver Operating Characteristic (ROC) curve (AUC) as shown in Figure 5.6 These metrics, extensively employed in geotechnical engineering studies (Hu et al. 2017; 2016), provided a comprehensive evaluation of model efficacy. Accuracy, a fundamental metric, measured the model's ability to correctly classify samples relative to the total sample size, ensuring an overall performance gauge that remained unbiased to sampling variations (Basheer et al. 2022). Recall and precision, intrinsically complementary, were balanced through the F1 score, a metric that considers both aspects while accommodating potential sampling biases. To further refine model assessment, the ROC curve was employed, aiding in the identification of optimal thresholds by plotting 1-Specificity against Sensitivity. The AUC, derived from the ROC curve, served as a comprehensive indicator of the model's overall discriminatory power, revealing its ability to distinguish between classes. Larger values across all these metrics signified superior model performance. Throughout the 3-step cross-validation test, average results for each metric were computed, ensuring a robust and representative evaluation of the models' classification capabilities. This systematic metric analysis, grounded in geotechnical engineering practices, thoroughly assesses the model's effectiveness in predicting groutability. Although confusion matrices are not traditionally used in geotechnical engineering, they play a critical role when applying machine learning classifiers such as those developed to distinguish groutable from non-groutable soils. The confusion matrix evaluates model performance by separating predictions into four categories: True Positives, where groutable soils are correctly identified; True Negatives, where non-groutable soils are correctly classified; False Positives, where non-groutable soils are incorrectly labelled as groutable, potentially leading to unnecessary or ineffective grouting efforts; and False Negatives, where groutable soils are mistakenly predicted as non-groutable, resulting in missed treatment opportunities. Beyond reporting overall accuracy, the matrix provides detailed insight into the types of errors the model makes, which is essential in engineering contexts where different misclassifications carry different practical risks. This structure enables the calculation of precision, recall, and F1-score, supporting the selection of reliable models for safety-critical decisions. By presenting the confusion matrix, the thesis offers geotechnical practitioners an intuitive and interpretable framework for understanding classification performance an important step as data-driven methods become increasingly integrated into geotechnical decision-making.

		Predictive label	
		Yes	No
True label	Yes	True Positve (*TP*)	False Negative (*FN*)
	No	False Positive (*FP*)	True Negative (*TN*)

INDEX	DEFINITION	FORMULA
Accuracy	Proportion of correctly classified samples	(TP+TN)/(TP+FN+FP+TN)
Precision	Proportion of correctly predicted positive Samples among all predicted positive samples	TP/(TP+FP)
Recall	Proportion of correctly predicted positive Samples among all actual positive samples	TP/(TP+FN)
F1	The harmonic mean of the precision and recall	2TP/(2TP+FN+FP)
Specificity	Proportion of correctly predicted negative Samples among all actual negative samples	TN/(FP+TN)
Sensitivity	Proportion of correctly predicted positive Samples among all actual positive samples	TP/(TP+FN)
AUC	The area under the curve (AUC) of the receiver operating characteristic (ROC)	

Fig. 5.6 Explanation of metrics and their formulas

5.6. Results

5.6.1 Performance comparison of the hybrid models and other methods

The performance of different models in predicting the groutability of soil was evaluated using a variety of metrics, including accuracy, precision, recall, F1-score, and the area under the curve (AUC). The results of testing data in Figure 5.7. indicate that the Hybrid BN emerges as the top-performing model, achieving an accuracy of 88.23%, precision of 84.2%, recall of 85.94%, F1-score of 85.06%, and AUC of 95.74%(or 0.9574) on the testing data.

This suggests the superiority of the Hybrid BN in accurately predicting groutability compared to the other models assessed, which include Support Vector Machine (SVM), Naïve Bayes (NB), and Discrete Bayesian Network.

Among the alternative models, the SVM model also demonstrates a commendable performance, particularly in precision (86.95%) and F1-score (86.44%), though with a comparatively lower AUC of 81.32% (or 0.8132) than the Hybrid BN. In contrast, the NB and Discrete BN models exhibit relatively inferior predictive capabilities across various metrics, with NB achieving an accuracy of 82.35% and the Discrete BN recording an AUC of 74.19% (0.7419), indicating their limitations in accurately classifying groutability. On the training data (Figure 5.8), the Hybrid BN sustains its robust performance, maintaining an accuracy of 80%, precision of 81.67%, recall of 82.63%, F1-score of 82.14%, and AUC of 88.53% (0.8853). The higher F1 scores achieved by the Hybrid BN suggest its ability to strike a better balance between precision and recall, effectively minimising both false positives and false negatives, resulting in more reliable classification. Additionally, the higher AUC values attained by the Hybrid BN signify its stronger discriminatory power in distinguishing between positive and negative instances across various thresholds, leading to fewer misclassifications and improved overall performance.

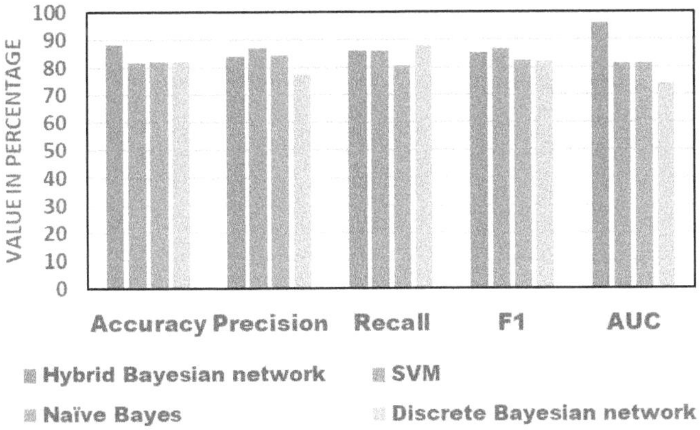

Fig 5.7. Performance of models on Testing data.

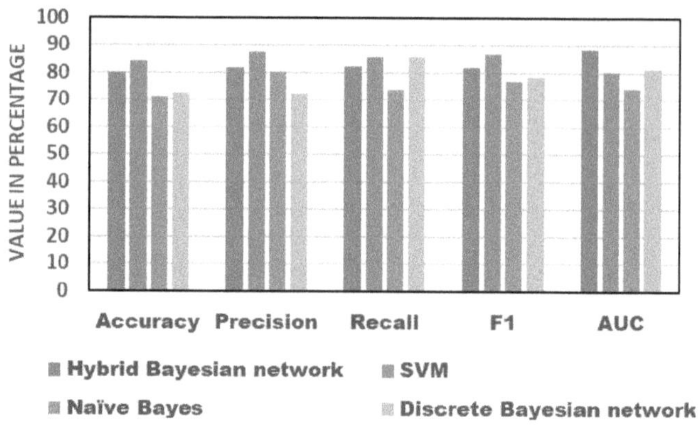

Fig 5.8. Performance of models on Training data.

5.6.2 Performance comparison on the Validation dataset

The validation set results are shown in Figure 5.9. demonstrate that the Hybrid BN outperforms other models in terms of accuracy, precision, recall, F1 score, and AUC. With an accuracy of 88%, Hybrid BN exhibits robust generalisation capabilities, suggesting its effectiveness in handling unseen data. This indicates that the Hybrid BN model effectively captures underlying patterns and relationships within the data, leading to reliable predictions across various scenarios. Compared to SVM, NB, and Discrete BN, Hybrid BN achieves higher performance across multiple evaluation metrics, showcasing its superiority in predictive power and model generalisation. Furthermore, Hybrid BN's ability to combine the strengths of Bayesian networks with hybrid structures enables it to efficiently model complex relationships in data, resulting in enhanced predictive performance and improved generalisation capabilities.

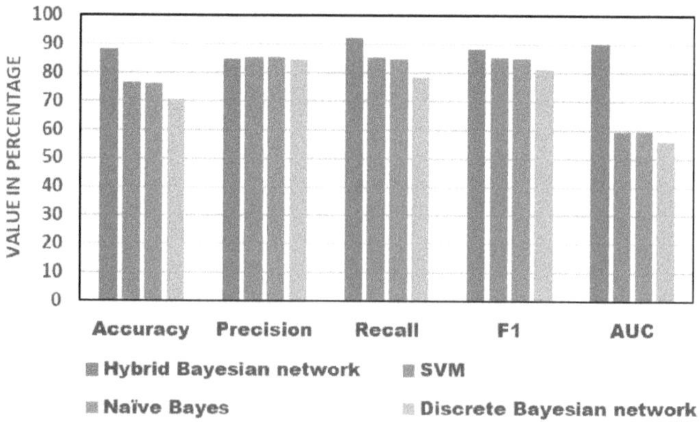

Fig 5.9. Performance of models on Validation data.

Employing a Hybrid BN model can significantly boost prediction accuracy while also improving computational efficiency. This model outperforms the Discrete BN model and other alternatives considered in the study. However, it's crucial to recognise limitations in the proposed Hybrid BN method. The assumption of a universal Gaussian distribution for continuous variables may not be universally applicable, as some phenomena may deviate from such a distribution. This limitation poses challenges to the widespread adoption of the Hybrid BN method, as this assumption leads to a significant loss of information regarding continuous variables.

5.7 Discussion

Bayesian Network (BN) methodology, incorporating both continuous and discrete variables, was applied to predict the groutability of soils. Continuous and Hybrid BN models were developed using Markov Chain Monte Carlo (MCMC) methods. The performance of these models was assessed by comparing them against discrete BN models and other existing approaches. The evaluation utilised three distinct arrangements of test datasets as well as a separate validation set. Compared to the Discrete BN model, the Hybrid BN framework integrates continuous variables directly, eliminating the need for data discretisation. This approach minimises the potential information loss typically associated with the discretisation process (Hu et al. 2022). In geotechnical applications, minimising information loss associated with discretisation ensures that the continuous variation of soil properties is accurately represented. Discretisation often oversimplifies these properties into predefined categories, potentially masking critical transitions or gradients. By avoiding discretisation, the hybrid approach retains the inherent variability and complexity of geotechnical data, leading to more precise modelling of soil behaviour and better prediction of parameters. This preservation of detail is crucial for designing reliable and efficient geotechnical systems. Parameter learning further underscores the computational advantages of continuous and Hybrid BN models. When learning parameters via equations such as Eq. (8) and Eq. (9), the storage requirements for a Discrete BN model with T nodes are represented as a $T \times L$ two-dimensional matrix, where L is the number of discrete states for each variable. In contrast, a continuous BN model with T nodes only requires a $T \times 2$ two-dimensional matrices, significantly reducing the computational burden. This reduced storage complexity not only enhances parameter learning efficiency but also contributes to faster model training and evaluation processes. Continuous and Hybrid BN models offer dual benefits over Discrete BN models: they improve prediction accuracy by preserving the natural variability of data and minimise computational inefficiencies through streamlined parameter storage and learning (Hu et al. 2022). In this study, no missing data were observed in the dataset, as all entries were complete and thoroughly preprocessed prior to model development. However, in practical scenarios involving field data, the presence of missing values is a common challenge. Bayesian networks (BNs) are inherently capable of addressing such issues due to their probabilistic framework, which allows for inference based on observed data while accounting for dependencies among variables. Additionally, BNs propagate uncertainty effectively, maintaining robust performance even in the presence of incomplete data.

The prediction of groutability is heavily dependent on maximising recall, as it is crucial to identify as many groutable soil samples as possible under varying soil and grout conditions. This emphasis on recall ensures that potentially groutable samples are not overlooked, even if it results in a higher rate of false positives. Consequently, the BN model tends to favour recall-focused predictions, which may skew its outcomes toward detecting groutable samples more effectively. This trade-off aligns with the primary objective of groutability analysis, where minimising the risk of missed groutable cases outweighs concerns regarding false positive identifications. The goal of groutability prediction is to correctly classify whether grouting will be successful (1) or unsuccessful (0). Since identifying successful cases is more important, the BN model is designed to focus on this, even if it slightly increases the chances of false positives. This means the model is biased toward ensuring groutable samples are not missed, which aligns with the practical needs of groutability prediction. Despite challenges such as class imbalance and sampling bias (evident from the groutable to non-groutable sample ratio of 1.29 in the dataset), the BN model demonstrated reduced sensitivity to these issues compared to alternative methods. Interestingly, the inherent class imbalance and sampling bias appeared to enhance the trained BN models' ability to predict groutability potential effectively for diverse soil samples. In model evaluation, it is essential to consider not only performance metrics but also aspects such as interpretability, computational efficiency, and the capacity for effective information extraction. Compared to alternative approaches, BN models demonstrate a robust theoretical foundation grounded in mathematical principles. Their structural learning process effectively integrates domain-specific knowledge and historical data, resulting in a model capable of probabilistic reasoning with high explanatory power. This combination of theoretical rigour and practical interpretability positions BN models as a compelling choice for complex predictive tasks. Furthermore, the BN method demonstrates the capability to perform predictions effectively even in scenarios characterised by data scarcity. Its flexibility extends to seamless model updates with the incorporation of new data, requiring no alterations to the established structure (Hu and Liu 2019). These attributes significantly enhance the practical applicability of BN models, making them highly advantageous in real-world scenarios where data limitations and dynamic environments are common. The proposed Hybrid BN method has limitations, such as the assumption of a Gaussian distribution for all continuous variables, even though some phenomena behave differently from the Gaussian distribution. Although the Gaussian or log-normal distribution is the most common in many geotechnical problems (Cao and Wang 2014). As the dataset contains many variables with values between 0 and 1, applying a log-normal distribution results in negative log-transformed values. This can distort the data distribution, complicating interpretation and potentially affecting the validity of statistical models. This is not conducive to the promotion and application of the Hybrid BN method, because the assumption would only lose a small amount of continuous variable information. If the performance of the Hybrid BN model needs improvement, the copula, which is a "distribution-generating" function, could be an alternative approach to characterise the inherent nonlinear correlation between variables with different marginal distributions. However, this approach cannot perform precise reasoning in the BN model, and it can only allow nodes to be continuous variables.

5.8 Conclusions

The study introduces a Hybrid Bayesian Network (BN) methodology for predicting groutability, representing a significant advancement in geotechnical modelling. The model's robust performance, validated through a three-step cross-validation process and application to a separate validation set, demonstrates its ability to generalise across diverse datasets effectively while mitigating overfitting concerns. Unlike traditional discrete BN models, the hybrid approach integrates both continuous and discrete variables, eliminating the need for discretisation and preserving the inherent variability of geotechnical data. This leads to improved accuracy, computational efficiency, and better modelling of soil behaviour.

Using Markov Chain Monte Carlo (MCMC) methods, the Hybrid BN outperforms Discrete BN models and other classification techniques like Support Vector Machines (SVM) and Naïve Bayes (NB). The ability to handle continuous variables without discretisation minimises information loss and streamlines parameter learning. These advancements align the model closely with practical engineering needs, emphasising recall to prioritise the identification of groutable soil samples, even under challenging conditions such as class imbalance and sampling bias. Moreover, the Hybrid BN model showcases flexibility in handling data scarcity and dynamic environments, with seamless updates enabled by its robust theoretical foundation and integration of domain knowledge. While this study relies on Gaussian assumptions, future research should explore copulas distribution Hybrid BN models to enhance versatility further. Implemented using the R language and the *bnlearn* package, this innovative approach significantly improves groutability prediction and sets the stage for future advancements in geotechnical engineering.

References

Afrazi M, Yazdani M. Determination of the effect of soil particle size distribution on the shear behaviour of sand. Journal of Advanced Engineering and Computation. 2021 Jun 30;5(2):125-34. http://dx.doi.org/10.25073/jaec.202152.331.

Agrawal, R. K., and Rajni Bala. "Incremental Bayesian classification for multivariate normal distribution data." Pattern Recognition Letters 29.13 (2008): 1873-1876.https://doi.org/10.1016/j.patrec.2008.06.010

Akbulut, Suat, and Ahmet Saglamer. "Estimating the groutability of granular soils: a new approach." Tunnelling and underground space technology 17.4 (2002): 371-380.https://doi.org/10.1016/S0886-7798(02)00040-8

Asadizadeh, Mostafa, and Abbas Majdi. "Developing new Adaptive Neuro-Fuzzy Inference System models to predict granular soil groutability." International Journal of Mining and Geo-Engineering 53.2 (2019): 133-142. https://doi.org/10.22059/ijmge.2018.255209.594728

Asghari, Kayvan, et al. "A fixed structure learning automata-based optimization algorithm for structure learning of Bayesian networks." Expert Systems 38.7 (2021): e12734.https://doi.org/10.1111/exsy.12734.

Avci, E. Groutability of Ultrafin 12 cement grout into sands at various relative density and gradation. Diss. Dissertation, Gazi University, 2009.

Basheer, Mohammed, et al. Machine Learning and Sensitivity Analysis Approach to Quantify Uncertainty in Landslide Susceptibility Mapping. World Bank, 2022.

Burwell Jr, Edward B. "Cement and clay grouting of foundations: practice of the corps of engineers." Journal of the Soil Mechanics and Foundations Division 84.1 (1958): 1551-1.https://doi.org/10.1061/JSFEAQ.0000099

Cabalar, Ali Firat, Abdulkadir Cevik, and Candan Gokceoglu. "Some applications of adaptive neuro-fuzzy inference system (ANFIS) in geotechnical engineering." Computers and Geotechnics 40 (2012): 14-33. https://doi.org/10.1016/j.compgeo.2011.09.008.

Cao, Zijun, and Yu Wang. "Bayesian model comparison and characterization of undrained shear strength." Journal of Geotechnical and Geoenvironmental Engineering 140.6 (2014): 04014018.https://doi.org/10.1061/(ASCE)GT.1943-5606.0001108.

Cheng, Min-Yuan, and Nhat-Duc Hoang. "Groutability estimation of grouting processes with microfine cements using an evolutionary instance-based learning approach." Journal of Computing in Civil Engineering 28.4 (2014): 04014014. https://doi.org/10.1061/(ASCE)CP.1943-5487.0000370.

Cheng, Min-Yuan, Nhat-Duc Hoang, and Yu-Wei Wu. "Hybrid intelligence approach based on LS-SVM and Differential Evolution for construction cost index estimation: A Taiwan case study." Automation in Construction 35 (2013): 306-313.https://doi.org/10.1016/j.autcon.2013.05.018

De Beer, E. E. "Grondmechanica, Deel II, Funderingen." Uitgeversmij NV Standard Boekhandel, Antwerp (1949): 45.

Gehl, Pierre, John Douglas, and Dina d'Ayala. "Inferring earthquake ground-motion fields with Bayesian networks." Bulletin of the Seismological Society of America 107.6 (2017): 2792-2808.https://doi.org/10.1785/0120170073.

Ghanizadeh, Ali Reza, et al. "Developing bearing capacity model for geogrid-reinforced stone columns improved soft clay utilizing MARS-EBS hybrid method." Transportation Geotechnics 38 (2023): 100906.https://doi.org/10.1016/j.trgeo.2022.100906

Glymour, Clark, Richard Scheines, and Peter Spirtes. Causation, prediction, and search. MIT press, 2001.

Hanif, Asif, Saad Ali, and Ali Ahmed. "A framework for fault diagnosis using continuous Bayesian network and causal inference." 2021 IEEE 19th International Conference on Industrial Informatics (INDIN). IEEE, 2021. https://doi.org/10.1109/INDIN45523.2021.9557490

Hu, Jilei, and Huabei Liu. "Bayesian network models for probabilistic evaluation of earthquake-induced liquefaction based on CPT and Vs databases." Engineering geology 254 (2019): 76-88. https://doi.org/10.1016/j.enggeo.2019.04.003

Hu, Jilei, et al. "Continuous-discrete hybrid Bayesian network models for predicting earthquake-induced liquefaction based on the Vs database." Computers & Geosciences 169 (2022): 105231.https://doi.org/10.1016/j.cageo.2022.105231

Hu, Ji-Lei, Xiao-Wei Tang, and Jiang-Nan Qiu. "Analysis of the influences of sampling bias and class imbalance on performances of probabilistic liquefaction models." International Journal of Geomechanics 17.6 (2017): 04016134.https://doi.org/10.1061/(ASCE)GM.1943- 5622.0000808.

Hu, Ji-Lei, Xiao-Wei Tang, and Jiang-Nan Qiu. "Assessment of seismic liquefaction potential based on Bayesian network constructed from domain knowledge and history data." Soil Dynamics and Earthquake Engineering 89 (2016): 49-60.https://doi.org/10.1016/j.soildyn.2016.07.007.

Huang, C. L., J. C. Fan, and W. J. Yang. "A study of applying microfine cement grout to sandy silt soil." Sino-geotech 111.1 (2007): 71-82.

Incecik, M., and I. Ceran. "Cement grouting model tests." Bulteni-Istanbul Teknik Universitesi 48 (1995): 305-318.

Kamran, Khalil Valizadeh, et al. "A comparative approach of support vector machine kernel functions for GIS-based landslide susceptibility mapping." Applied Geomatics 13.4 (2021): 837-851.https://doi.org/10.1007/s12518-021-00393-0

Kim, Jong-Sun, et al. "Groutability of cement-based grout with consideration of viscosity and filtration phenomenon." International Journal for Numerical and Analytical Methods in Geomechanics 33.16 (2009): 1771-1797. https://doi.org/10.1002/nag.785.

Kononenko, Igor. "Inductive and Bayesian learning in medical diagnosis." Applied Artificial Intelligence an International Journal 7.4 (1993): 317-337.https://doi.org/10.1080/08839519308949993.

Landry, Eric, Daniel Lees, and Alex Naudts. "New developments in rock and soil grouting: design and evaluation." GEOTECHNICAL NEWS-VANCOUVER- 18.3 (2000): 38-48.

Maldonado, A. D., P. A. Aguilera, and A. Salmerón. "Continuous Bayesian networks for probabilistic environmental risk mapping." Stochastic environmental research and risk assessment 30 (2016): 1441-1455.https://doi.org/10.1007/s00477-015-1133-2.

Morales-Nápoles, Oswaldo, et al. "A continuous Bayesian network for earth dams' risk assessment: methodology and quantification." Structure and Infrastructure Engineering 10.5 (2014): 589-603.https://doi.org/10.1080/ 15732479.2012.757789.

Movassagh, Ali Akbar, et al. "Artificial neural networks training algorithm integrating invasive weed optimization with differential evolutionary model." Journal of Ambient Intelligence and Humanized Computing (2023): 1-9. https://doi.org/10.1007/s12652-020-02623-6.

Nowroozi, Vahid, et al. "Optimum design for soil nailing to stabilize retaining walls using FLAC3D." Journal of Advanced Engineering and Computation 5.2 (2021): 108-124.http://dx.doi.org/10.25073/jaec.202152.329

REJWANUR RAHMAN, M. D., et al. "PROBABILISTICS SOIL CHARACTERIZATION BASED ON PORTABLE FREE FALL PENETROMETER MEASUREMENTS." Coastal Sediments 2023: The Proceedings of the Coastal Sediments 2023. 2023. 2398-2407. https://doi.org/10.1142/9789811275135_0220

Sang-To, Thanh, et al. "A new metaheuristic algorithm: Shrimp and Goby association search algorithm and its application for damage identification in large-scale and complex structures." Advances in Engineering Software 176 (2023): 103363.https://doi.org/10.1016/j.advengsoft.2022.103363.

Scutari, Marco. "Learning Bayesian networks with the bnlearn R package." Journal of statistical software 35 (2010): 1-22.

Sebastiani, Fabrizio. "Machine learning in automated text categorization." ACM computing surveys (CSUR) 34.1 (2002): 1-47. https://doi.org/10.1145/505282.505283.

Tekin, E. "Experimental studies on the groutability of microfine cement (Rheocem 900) grouts to sands having various gradations." Gazi Univ Google Sch (2004).

Tekin, E. R. H. A. N., and S. O. Akbas. "Artificial neural networks approach for estimating the groutability of granular soils with cement-based grouts." Bulletin of Engineering Geology and the Environment 70 (2011): 153-161. https://doi.org/10.1007/s10064-010-0295-x.

To, Thanh Sang, et al. "A nonlinear optimization method for calibration of large-scale deep cement mixing in very soft clay deep excavation." International Journal for Numerical and Analytical Methods in Geomechanics 48.8 (2024): 1949-1978. https://doi.org/10.1002/nag.3714.

Torres-Torriti, Miguel, and Felipe Calderara Cea. "Statistical machine learning." Encyclopedia of Smart Agriculture Technologies. Cham: Springer International Publishing, 2023. 1-32. https://doi.org/10.1007/978-3-030-89123-7_227-1.

Zebovitz, S., R. J. Krizek, and D. K. Atmatzidis. "Injection of fine sands with very fine cement grout." Journal of geotechnical engineering 115.12 (1989): 1717-1733. https://doi.org/10.1061/(ASCE)0733-9410(1989)115:12(1717).

CHAPTER

6

Bayesian Logistic Regression for Soil Liquefaction Using CPT Data

6.1 Introduction

In geotechnical engineering, geotechnical models such as soil liquefaction and bearing capacity models have been widely used to predict the performance of geotechnical systems. However, like any model, it is just a duplication of the real world in a simplified way, as some model uncertainty always exists (Ang and Tang 1984; Cheung and Tang 2005). Hazard by liquefaction (e.g., settlement, sand boiling, lateral expansion, and ground cracking) causes significant damage to the infrastructure during earthquakes (Oommen 2010; Kohestani et al.2015). Therefore, assessing the occurrence of soil liquefaction potential and its impact in a seismically active region is essential for seismic hazard mitigation. Due to several related factors' complexity, uncertainties, and randomness, predicting the seismic liquefaction potential is a probabilistic rather than a deterministic problem. It is often unable to meet seismic hazard analysis requirements in engineering. This becomes a major cause for the prediction of liquefaction in the soil concerning the geotechnical property (Bhardwaj et al. 2022). Therefore, predictive seismic liquefaction methods based on probabilistic correlations may better meet technical requirements. Various empirical and semi-empirical methods for quantifying liquefaction are presented in the literature (Youd 2001). In recent times rapid development of computer technology and mathematics has led to the development of many new methods based on probabilistic correlations to assess seismic liquefaction, including the artificial neural network (ANN) method (Anthony 1994; Anthony 1996), the support vector machine (SVM) method (Anthony 2007; Pal 2006) and the reliability method (Jha and Suzuki 2009). However, existing methods of artificial intelligence have the following shortcomings (Liang et al 2012) (1) limited use of prior knowledge makes interpreting assessment results difficult; (2) multiple sources of information cannot be integrated into one model; and (3) require a large amount of data to create a correct model. The mathematical concept used in this study, i.e., Bayesian statistics, is a rapidly growing field with geotechnical safety and reliability (Carlin & Louis 2009; Gelman et al. 2013; Givens and Hoeting 2013). Bayesian analysis provides an effective tool to improve understanding of uncertain soil properties by combining prior knowledge with limited site-specific information (Straub 2015; Jiang et al. 2017; Zhang et al. 2017). Bayesian methods are widely accepted for estimating the uncertainties that are coalesced within the model. (Sherri et al.2021; Gelfand and Smith 1990) introduced MCMC methods for sampling from the posterior distribution in logistic regression when conjugate priors do not exist. These methods, pioneered by statistical physicists in the 1950s, involve constructing a Markov chain with a stationary distribution equal to the target posterior. The Metropolis-Hastings (MH) algorithm, initially proposed by Metropolis et al. 1953 and generalised by Hastings 1970, is a key algorithm in this regard. It uses an accept-reject strategy to generate the Markov chain and has sparked renewed interest in Bayesian statistics with the advent of powerful computers. While the MH algorithm is widely used in Bayesian logistic regression, the Gibbs sampler can also be employed if sampling from conditional distributions is feasible. Confidence bounds based on normality assumptions are typically used for parameter estimation uncertainty in frequentist logistic regression, whereas Bayesian logistic regression provides probabilistic estimates through simulation methods like MCMC. Logistic regression has been extensively applied in landslide susceptibility studies worldwide, with various researchers utilizing logistic regression models (Dai and Lee 2003; Lee 2004; 2005; Ayalew and Yamagishi 2005; Lee and Pradhan 2007). While numerous approaches are available for evaluating liquefaction potential, many rely on simplified empirical models that often lack a solid statistical foundation. Established techniques, like the Seed and Idriss method, use in-situ data such as SPT and CPT but typically involve indirect relationships or assumptions. CPT-based methods have gained popularity for their ability to provide continuous and detailed soil profiles; however, most empirical correlations for estimating liquefaction do not undergo thorough statistical validation. Logistic regression models developed for probabilistic liquefaction assessment frequently overlook essential statistical assumptions such as variable independence, normal distribution, and linearity with the logit function. Ignoring these assumptions can compromise the reliability and consistency of the models. Additionally, previous research has seldom confirmed whether their explanatory variables meet these basic requirements. As a result, model outcomes may differ depending on how inputs are formulated. There remains a clear need to create CPT-based liquefaction models that are both statistically rigorous and widely applicable. This calls for integrating sound statistical analysis with geotechnical principles. A well-validated, probabilistic approach would significantly improve the reliability of liquefaction predictions.

The need to perform a Bayesian logistic regression analysis of field yield data to generate empirical equations for interpreting liquefaction probability is immense. A curve developed by (Liao et al.1988) won considerable attention as a tool to characterise the liquefaction called the limit curve. The curve that separates liquefaction from non-liquefaction in the simplified methods. Toprak et al. (1999) performed a logistic regression analysis of field records using the curve.

This study presents a Bayesian logistic regression approach to assess the probability of seismic-induced liquefaction based on cone penetration test (CPT) data. The primary objective is to formulate a probabilistic model that distinguishes between liquefied and non-liquefied cases using statistically selected input variables derived from rigorous probability-based evaluations. The model is developed using a comprehensive database of 399 field case histories from Taiwan, China, and the United States. Soil types are classified solely on CPT data, and separate regression equations are established for each class following detailed statistical analysis. The resulting logistic regression formulations are expressed in a structured mathematical system, enhancing their applicability to similar binary classification problems in geotechnical engineering. To validate the effectiveness of the proposed model, comparative evaluations are performed against a logistic regression model from the literature, thereby highlighting the advantages of the Bayesian framework in terms of uncertainty quantification and model robustness.

6.2. Bayesian logistic regression

6.2.1 Logistic regression

Linear regression is typically applied when the response variable is continuous and is modelled as a function of one or more explanatory variables or covariates. However, when the response variable is binary, taking only the values 0 or 1, it is still possible to fit a linear regression model. Figure 6.1 illustrates this case, where the explanatory variable is represented on the x-axis and the binary response variable y is shown on the y-axis. Since y can only assume values of 0 and 1, the fitted regression line passes through these discrete points, as depicted in Figure 6.1.

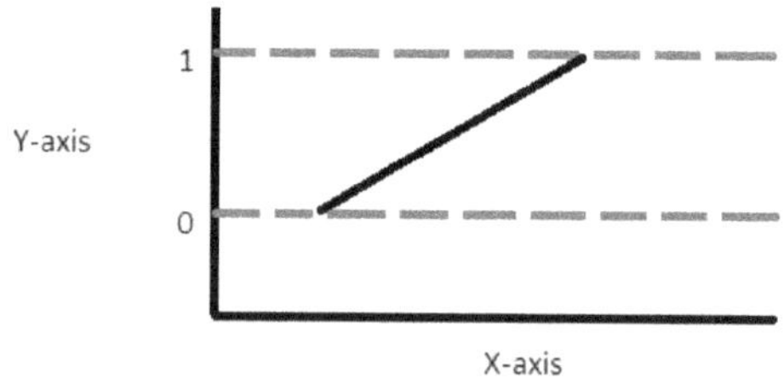

Fig 6.1 Representation of linear regression

As observed in the graph, only two particular values of x will cause a prediction of 0 or 1. Not to mention, if a residual plot from this regression is drawn, it would certainly not appear to be very normally distributed. Instead of using a normal likelihood, it is more appropriate to use a Bernoulli likelihood here.

$$y_{i/\phi_i} \text{ is independent of the Bernoulli distribution}(\phi_i) \tag{6.1}$$

i=1,2,3,4

That is, the response y_i, given ϕ_i, would be independent of a Bernoulli distribution, where ϕ_i is the probability of success for observation i. In this case, ϕ_i is also the expected value of y_i. But there's a need to relate this ϕ_i to our explanatory variable x. In linear regression, the expected value of y_i was equal to

$$E(y) = B_0 + B_i x_i \tag{6.2}$$

One option in this model would be to model ϕ_i directly with this linear form. But there's one problem: because ϕ_i represents a probability, it needs to lie between 0 and 1. One needs to place some pretty odd restrictions on the coefficients to keep this linear form right here between 0 and 1. Instead, a link function that relates to the linear form of the restricted parameter, in this case, ϕ_i, without any restrictions on the coefficients (B). To do this, the odds, which are defined as ϕ_i over 1 minus ϕ_i, are used. Odds can take on any value greater than 0. So, they are less restricted but not unrestricted. So, the log of the odds is,

$$\text{Log}\left(\frac{\phi}{1-\phi}\right) = B_0 + B_i x_i \tag{6.3}$$

This quantity is called the logit link, or logistic link. It can now take on any real number, positive or negative, so that it can be related to our linear form without any restrictions. That is, the logit of ϕ_i, which is equal to the log of the odds, will be equal to this linear form here. So, this is the piece that relates the x to the success probability. After solving the equation, the expression for the expected value of y_i can be recovered. From this, the expected value of y_i can be obtained by solving this equation here for ϕ_i,

$$E(y_i) = \phi_i = \frac{e^{B_0 + B_i x_i}}{1 + e^{B_0 + B_i x_i}} = \frac{1}{1 + e^{-(B_0 + B_i x_i)}} \tag{6.4}$$

This form is a little bit more computationally convenient. This process is called inverting the link function to get an expression for the expected value of y. If the expected value of the response variable is plotted as a function of x, the resulting curve would resemble the figure shown in Figure 6.2. Where the expected value now stays between 0 and 1. This is logistic regression using the logistic link. If the B coefficient for the covariate x is positive, it means that an increase in the value of x corresponds to an increase in the probability that y is 1. So, the curve would look like Figure 6.2. If the coefficient is negative, i.e., the B coefficient is negative, which means that an increase in x would correspond to a decrease in the probability, which is equal to 1.

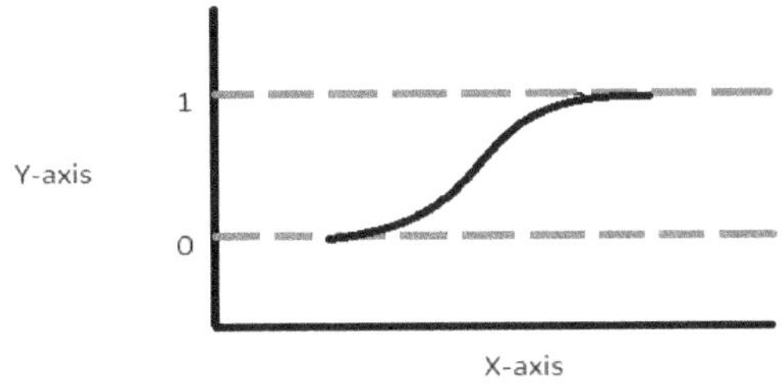

Fig 6.2 Representation of logistic regression

6.2.2 Bayesian Logistic Regression

Bayesian Logistic Regression (BLR) is a probabilistic framework for binary classification that models the posterior distribution of regression coefficients rather than point estimates. Unlike classical logistic regression, which provides fixed coefficient values via maximum likelihood estimation (MLE), BLR incorporates prior beliefs about parameters and uses observed data to update these beliefs through Bayes' theorem.

i) Model Formulation

Let $y_i \in \{0,1\}$denote the binary response variable and $x_i = (x_{i1}, x_{i2})$ denote the input feature vector for observation i. The logistic regression model defines the probability of class membership as:

$$P(y_i = 1 \mid x_i, B) = \sigma(\eta_i) = \frac{1}{1 + e^{-\eta_i}} \qquad (6.5)$$

where $\eta_i = B_0 + B_1 x_{i1} + B_2 x_{i2}$ and $B = (B_0, B_1, B_2)$ are the regression coefficients. The function σ(·) denotes the logistic sigmoid.

ii) Prior Distributions

In the Bayesian framework, prior distributions encode prior beliefs about the model parameters. Here, weakly informative Gaussian priors to all coefficients. These priors reflect the assumption that the coefficients are likely centred around zero but with considerable uncertainty.

$$B_j \sim N(0, \sigma^2) \qquad (6.5)$$

with σ =10 for j=0,1,2

iii) Likelihood

Assuming conditional independence, the likelihood of observing the binary outcomes y (y1,...,yn) given the covariates and parameters is modelled using the Bernoulli distribution:

$$y_i \sim Bernoulli(p_i) \qquad (6.6)$$

iv) Posterior Inference using MCMC

Bayesian inference proceeds by computing the posterior distribution over the parameters:

$$p(B \mid y, X) \propto p(y \mid X, B) \cdot p(B) \qquad (6.7)$$

Since this posterior is analytically intractable due to the non-conjugate likelihood, Approximation is done using Markov Chain Monte Carlo (MCMC) methods. Specifically, we employ the No-U-Turn Sampler (NUTS), an efficient gradient-based MCMC algorithm that adapts Hamiltonian Monte Carlo (HMC) by automatically tuning the trajectory length to avoid inefficient random walks. Sampling was conducted with 1000 posterior draws after a burn-in (tuning) period of 500 iterations, using a target acceptance probability of 0.90. The PyMC implementation is used in Python.

6.3 Conditions assumed for explanatory variables

In Bayesian logistic regression, the explanatory variables, typically denoted as X (X_1,X_2,…,X_k), are subject to several statistical assumptions to ensure valid and interpretable model inference. These assumptions are rooted in the theory of generalised linear models and are essential for the proper functioning of both classical and Bayesian logistic regression frameworks.

Let π(X)=P(Y=1|X) represent the predicted probability of the binary outcome given input features, and let logit (π(X)) =log($\frac{\pi(X)}{1-\pi(X)}$)denote its log-odds transformation.

According to Anderson (2003) and Johnson and Wichern (2002), the following presumptions should ideally hold for the explanatory variables:

(i) Independence Among Predictors.

Each explanatory variable X_k should be statistically independent of the others (i.e., low multicollinearity). This ensures that each predictor contributes unique information to the model.

(ii) Normality of the Predictor Distribution-The variables X_k should be approximately normally distributed. While Bayesian logistic regression does not strictly require normality (unlike linear regression), this assumption improves stability and interpretability in small-sample or inferential contexts.

(iii) Linearity in the Logit-The relationship between the logit-transformed outcome logit (π(X)) and the explanatory variables X_k should be linear as in Equation 6.8.

$$logit(\pi(X)) = B_0 + B_1X_1 + B_2X_2 \cdots + B_kX_k \quad (6.8)$$

Garson (2003) argued that the predicted probabilities π(X) themselves are not required to follow a normal distribution; instead, it is sufficient for them to belong to the exponential family of distributions (e.g., binomial, Poisson, normal, gamma). This aligns with the theoretical structure of generalised linear models (GLMs), to which logistic regression belongs. Johnson and Wichern (2002) further clarify that although π(X) may deviate from normality, the logit transformation logit(π(X)) must maintain a linear association with X to uphold the assumptions of the logistic model. Menard (2002) similarly emphasised that this linearity in the logit is fundamental for valid coefficient interpretation and estimation.

6.4 Data

The Bayesian logistic regression analysis in this study was conducted on a dataset comprising 399 observations, representing both liquefied and non-liquefied sites. Each record corresponds to a soil profile subjected to earthquake-induced loading, characterised by cone penetration test (CPT) parameters and liquefaction response.

The dataset integrates field case histories compiled from multiple seismic events:

i. Chi-Chi Earthquake (Taiwan, 1999): A total of 174 records were obtained from the study by Lai et al. (2004), which includes detailed CPT-based evaluations of liquefaction occurrence in sandy and silty soils.

ii. Other Global Earthquakes: An additional 225 cases were sourced from the compilation by Juang et al. (2000), which covers a diverse range of earthquakes, including:

- San Fernando Earthquake (California, 1971)
- Haicheng Earthquake (China, 1975)
- Imperial Valley Earthquakes (California, 1979 and 1981)
- Loma Prieta Earthquake (California, 1989)

For each site, key input variables such as the normalised cone tip resistance qc1N, cyclic stress ratio (CSR), and other geotechnical parameters were compiled based on standard CPT interpretations. The normalised resistance qc1N serves as an index for soil density and strength, while CSR quantifies the seismic demand at the site, standardised to a magnitude 7.5 event (CSR7.5). These parameters are widely recognised in empirical liquefaction assessments.

6.4.1 Variable Selection

In liquefaction modelling using Bayesian logistic regression, several input variables have been proposed in prior studies based on both deterministic frameworks and statistical classification techniques. These explanatory variables typically fall into two primary categories: first, those characterise the inherent resistance of soils to liquefaction, commonly represented by the cyclic resistance ratio (CRR); and second, those that quantify the level of seismic demand, typically represented by the equivalent cyclic stress ratio (CSR). The CSR is frequently evaluated through in situ testing techniques such as the Standard Penetration Test (SPT), Cone Penetration Test (CPT), and shear wave velocity measurements. In this study, using the CPT-based empirical procedure developed by Seed and Idriss (1971), which can be expressed as the dimensionless ratio $\frac{\sigma v}{\sigma v'}$ and calculated from the following equation in 6.9.

$$CSR7.5 = 0.65 * \frac{gd}{MSF} * \frac{a_{max}}{g} * \frac{\sigma v}{\sigma v'} \tag{6.9}$$

where depth $K \leq 9.5$

$$gd = 1.0 - 0.00765K$$

$9.15 < K \leq 239.15$

$$gd = 1.174 - 0.0267K$$

where σv and σv′ are the total and effective vertical overburden stresses in kg/cm^2, a_{max} is the peak horizontal ground acceleration, g is the gravitational acceleration, respectively, gd is the stress reduction coefficient, and MSF is the magnitude scaling factor. As proposed by

Liao et al. (1988), the average value of the depth reduction factor (gd) can be estimated as a function of depth below the ground surface (K). For depths less than or equal to 9.15 meters, the relationship is given by gd=1.0−0.00765*K, while for depths greater than 9.15 meters but not exceeding 23 meters, the equation becomes gd=1.174−0.0267*K. These empirical formulas are used to account for the reduction in seismic shear stress with increasing depth.

Idriss (1999) provided expressions to calculate the magnitude scaling factor (MSF), which adjusts for the effects of earthquake magnitude (Mw) on liquefaction triggering. When the earthquake magnitude is equal to or greater than 5.75, the MSF is defined by the equation MSF=37.9 $(Mw)^{-1.81}$. For magnitudes below 5.75, a constant value of MSF=1.625 is recommended. These relationships are essential for normalising seismic demand across different magnitudes and depths when assessing liquefaction potential.

In Cone Penetration Testing (CPT), the primary indicator of a soil's liquefaction resistance is the cone tip resistance (qc). However, because qc naturally increases with higher effective overburden stress, a correction must be applied to ensure consistency across varying depths. To address this, an overburden stress normalisation factor, CN, is introduced to adjust the raw qc values. This adjustment aligns the data with a standardised reference stress condition, typically equivalent to 100 kPa or 1 atmosphere. The normalisation approach adopted in equation 6.10 is based on the method proposed by Liao and Whitman (1986) for Standard Penetration Test (SPT) N-values, which has been extended for use with CPT data to allow for more accurate empirical correlations related to liquefaction resistance.

$$qc1N = C_N * \frac{qc}{Pa} = \sqrt{\frac{Pa}{\sigma v'}} * \frac{qc}{Pa} \tag{6.10}$$

For soil to be liquefied, 1 is used, and for soil not to be liquefied, 0 is used.

6.5. Results

6.5.1 Normality of the selected variables

To meet the assumptions of linearity and normality in logistic regression, the variables qc1Nq and CSR7.5 from the liquefaction dataset are transformed into alternative forms. Appropriate transformations are identified through probability plots, which compare the cumulative probability against each variable to assess its distribution. These plots are used to evaluate both the original and transformed variables based on 399 field liquefaction case histories, ensuring their suitability for regression modelling.

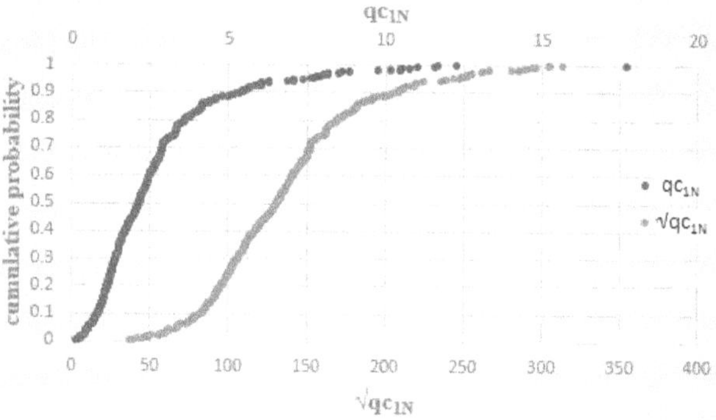

Fig 6.3. The cumulative probability distribution plot of cone penetration resistance (qc1N) and (√qc1N)

Figure 6.3 presents the cumulative probability distributions of the original cone penetration resistance (qc1N) and its square root-transformed counterpart (√qc1N) based on 399 field liquefaction case histories. In this figure, the blue markers represent the distribution of qc1N, while the red markers correspond to (√qc1N). The distribution of (√qc1N) more closely resembles the S-shaped form of a normal cumulative distribution function, whereas the qc1N data display a skewed pattern, with a steeper initial slope and a longer tail, indicating right-skewness. This visual comparison suggests that the square root transformation effectively reduces skewness and brings the distribution closer to normality. Consequently, (√qc1N) is considered more suitable for use in Bayesian logistic regression models that assume normally distributed explanatory variables. The skewness results show that the qc1N variable has a strong positive skewness (2.39), indicating a highly right-tailed distribution where most values lie on the lower side while a smaller proportion of large values stretch the tail. After applying a square-root transformation(√qc1N), the skewness decreases to 1.09, reflecting a notable improvement toward symmetry. Although still positively skewed, the transformed distribution is substantially less extreme, demonstrating that the square-root transformation effectively reduces the influence of high qc1N values. This transformation helps improve the suitability of the data for statistical modelling and regression analysis by making the distribution more balanced and less dominated by outliers.

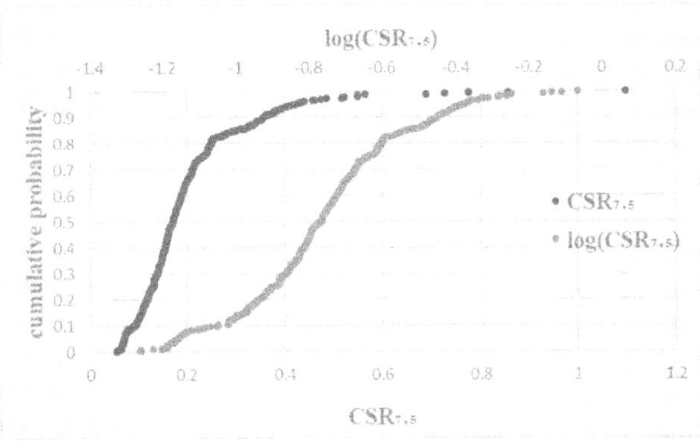

Fig 6.4 The cumulative probability distribution plot of (CSR7.5) and (log(CSR7.5)

Figure 6.4 displays the cumulative distributions of the cyclic stress ratio (CSR7.5) and its logarithmic transformation (log(CSR7.5) across 399 field liquefaction case histories. The red markers represent the original CSR7.5, while the blue markers correspond to the log(CSR7.5) values. The distribution of CSR7.5 exhibits a strong right-skew, as indicated by the sharp initial rise and the pronounced flattening in the upper cumulative probability range. In contrast, the logarithmically transformed values resemble the S-shaped form, which is more consistent with a normal distribution. This transformation effectively reduces skewness and improves the normality of the data. Therefore, log(CSR7.5) is preferred over the untransformed variable for inclusion in statistical models that assume normally distributed explanatory variables.

Moreover, higher-order terms such as $(\sqrt{qc1N})^k$ and $(\ln(CSR7.5))^k$ for k>1were excluded to maintain the assumption of linearity between the predictors and the logit function π(X). As a result, $\sqrt{qc1N}$ and ln(CSR7.5) were identified as the most appropriate explanatory variables for subsequent modelling. The skewness analysis indicates that the CSR7.5 variable exhibits a strong positive skewness (2.56), demonstrating a highly asymmetric distribution with a pronounced right tail. This suggests that a majority of the CSR7.5 values are clustered toward the lower range, while a smaller number of larger values extend the distribution to the right. In contrast, the log-transformed CSR7.5 shows a substantially reduced skewness of 0.29, indicating a distribution that is much closer to symmetry. This confirms that the logarithmic transformation is effective in stabilizing variance and reducing the influence of extreme values, thereby improving the suitability of the data for subsequent statistical modelling and regression-based analyses. Implementing the results of the probability examinations, the liquefaction probability can be explicitly expressed in equation 6.11.

$$PL = \frac{1}{1 + e^{-(B_0 + B_1 * \ln(CSR7.5)) + B_2 * (\sqrt{qc1N})}} \quad (6.11)$$

where *PL*=liquefaction probability for a site with $\sqrt{qc1N}$ and CSR7.5 and B_0, B_1, and B_2=regression coefficients

6.5.2 Liquefaction Probability Models

On performing Bayesian logistic regression using variables $\sqrt{qc1N}$, aan CSR7.5, and equation 6.8 to determine B_0, B_1, and B_2 from 399 data cases, equation 6.12 was found.

$$\pi(X) = (0.418 + 1.627 * \ln(CSR7.5)) - 1.181 * (\sqrt{qc1N}) \qquad (6.12)$$

Where $PL = \frac{1}{1+e^{-\pi(X)}}$

The posterior density plots of the model parameters (B_0, B_1, B_2), obtained through Bayesian logistic regression, are presented in Figure 6.5. These distributions quantify the uncertainty associated with each parameter estimate, offering more informative insights than single-point estimates typical of frequentist approaches. The spread of each distribution reflects the degree of confidence in the corresponding parameter, with narrower distributions indicating greater certainty. This probabilistic characterisation of the parameters enables rigorous uncertainty quantification, which is essential for assessing the reliability of liquefaction predictions under varying soil and seismic conditions. As shown in Figure 6.5, the orange and blue curves represent the kernel density estimates of the sampled values from each chain. The y-axis, which would typically represent the density values, the primary objective is to visualise the shape, location, and overlap of the posterior distributions. Wider distributions indicate that the model acknowledges a broader range of plausible values, which is particularly valuable when data are noisy or limited. Instead of overstating confidence, the Bayesian approach transparently communicates uncertainty and propagates it into predictions. This wide coverage of uncertainty makes the model more reliable for prediction, as it accounts for variability in parameter estimates and avoids overfitting to a narrow set of values.

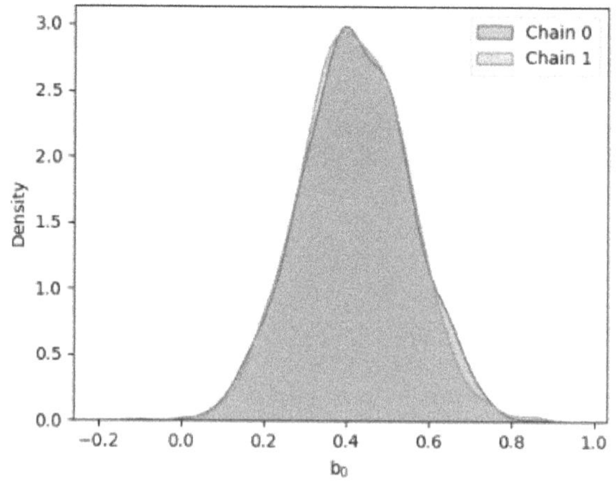

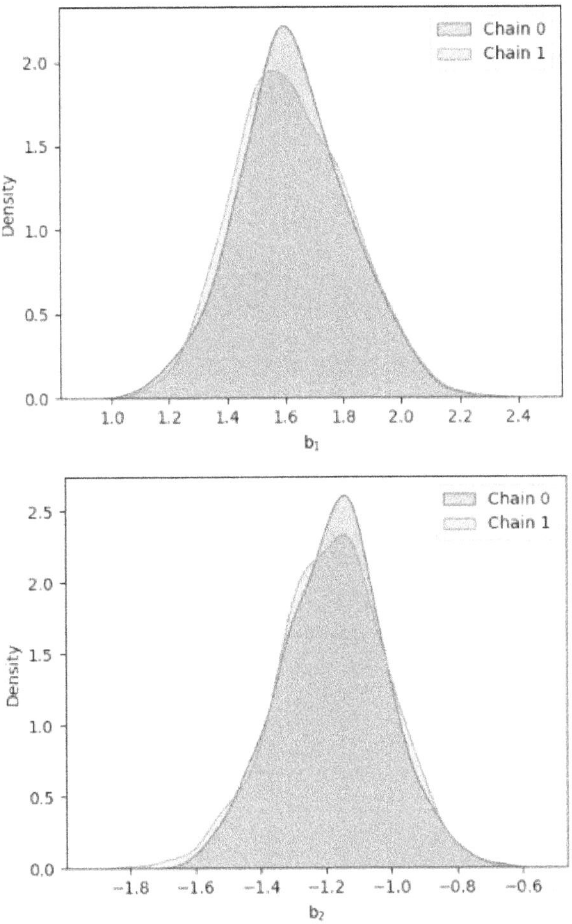

Fig 6.5. Posterior density plots of the coefficients

6.5.3 Liquefaction Probability Models for Different Soil Types

In approaches for assessing liquefaction triggering, the type of soil plays a vital role in influencing its cyclic resistance. Since Cone Penetration Testing (CPT) does not retrieve physical soil samples, soil classification must be inferred from test data. This study uses the Soil Behaviour Type Index (Ic), a parameter widely recognised in engineering practice, to classify soils based on CPT results. The index Ic is calculated using the modified equation 6.13 from Robertson (1990):

$$\mathrm{Ic} = [(3.47 - \log(\mathrm{qc1N}))^2 + (\log \mathrm{F} + 1.22)^2]^{0.5} \qquad (6.13)$$

Here, F, the normalised friction ratio, is given by:

$$F = \frac{f_s}{q_c - \sigma_v} \tag{6.14}$$

Where f_s is the sleeve friction in kg/cm^2.

Based on the computed Ic values, soils are grouped into three categories:

- Sand: $1.31 \leq Ic \leq 2.05$
- Sand mixture: $2.05 < Ic \leq 2.60$
- Silt mixture: $2.60 < Ic \leq 2.95$

To ensure consistency with this classification, records with Ic<1.31 or Ic>2.95 are excluded, resulting in the removal of 3 cases. The remaining 396 cases were categorised accordingly, and separate Bayesian logistic regression models were developed for each soil type to estimate the probability of liquefaction occurrence. The model formulations for each soil class are provided in Table 6.1.

Table 6.1 Bayesian Logistic Regression Probability Models for Different Soil Types

Soil classification	Logit transformation of liquefaction probability	Number of cases
Sand	$\pi(X) = (-0.253 + 1.730 * \ln(CSR7.5)) - 1.201 * (\sqrt{qc1N})$	198
Sand Mixture	$\pi(X) = (0.961 + 1.636 * \ln(CSR7.5)) - 0.323 * (\sqrt{qc1N})$	181
Silt Mixture	$\pi(X) = (1.970 + 2.605 * \ln(CSR7.5)) + 0.121 * (\sqrt{qc1N})$	17

Where $PL = \frac{1}{1+e^{-\pi(X)}}$

1st equation in Table 6.1 is used to plot the contours in Figure 6.6. The different probability demonstrates a nonlinear interaction between the seismic demand (CSR7.5) and soil resistance (qc1N) in predicting the probability of liquefaction for sand. While both predictors exhibit positive associations with liquefaction probability, CSR7.5 increases the likelihood due to greater seismic loading, and qc1N reflects higher soil resistance. The set of probability contours (P = 0.1 to 0.95) visually partitions the qc1N–CSR7.5 space into zones of increasing liquefaction probability, with the 0.5 contour (green) often used as the threshold for binary classification.

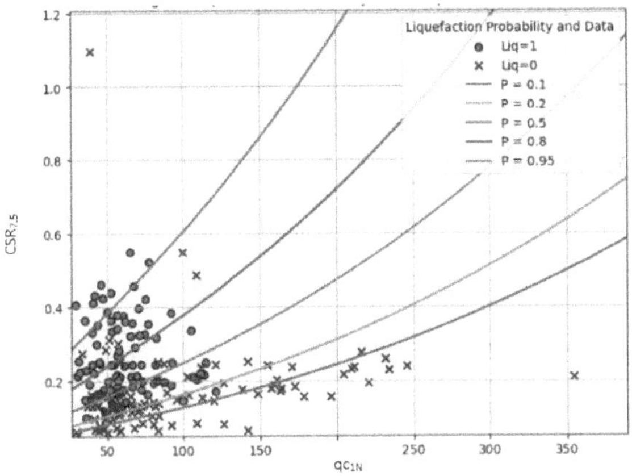

Fig 6.6 Bayesian logistic regression soil model for Sand (1.31≤Ic≤2.05)

The 2nd equation in Table 6.1 is used to plot the contours in Figure 6.7. For sand mixtures. The contours suggest a more subdued and nearly linear relationship between the explanatory variables and the probability of liquefaction, particularly when compared to pure sands. The coefficient of √qc1N is relatively small, implying a weaker influence of soil resistance on liquefaction probability, and the spacing of the probability contours (P = 0.1 to 0.95) appears more uniform and parallel in the plot, reinforcing the quasi-linear behaviour. This linear-like structure indicates that sand mixtures exhibit more gradual transitions in liquefaction susceptibility with increasing qc1N or CSR7.5.

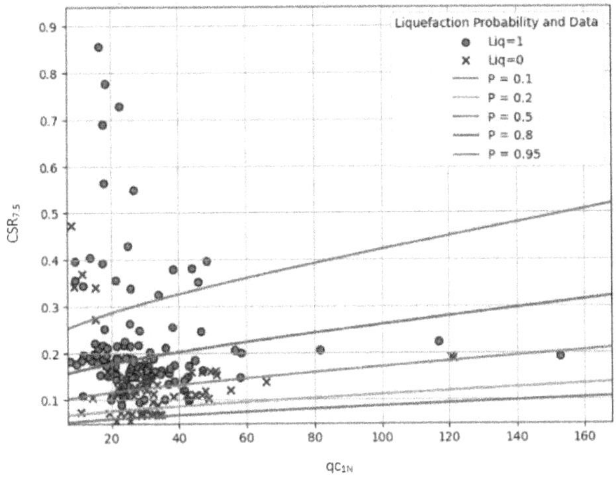

Fig 6.7 Bayesian logistic regression soil model for Sand Mixture (2.05<Ic≤2.60)

The 3nd equation in Table 6.1 is used to plot the contours in Figure 6.8 for the silt Mixture. Unlike sand and sand mixtures, this equation suggests a positive association of both CSR7.5 and qc1N with the probability of liquefaction. Interestingly, the coefficient for √qc1N is positive, which is counterintuitive since higher tip resistance usually implies greater liquefaction resistance. This anomaly may be attributed to the limited sample size or potentially non-representative data distribution in the silt mixture dataset. The plotted probability contours appear relatively flat and exhibit minimal curvature, which might give the illusion of linearity; however, this could stem from the sparsity and clustering of data points. Given the small range and low number of observations, especially in the upper qc1N values, the model may not capture the true nonlinear behaviour typically observed in liquefaction phenomena.

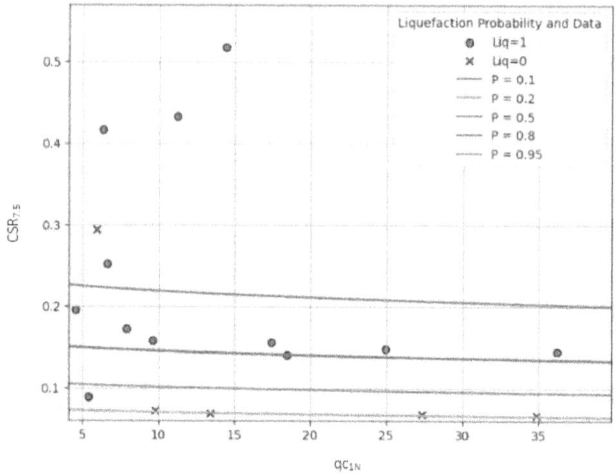

Fig 6.8 Bayesian logistic regression soil model for Silt mixture (2.60<Ic≤2.95)

These curves represent the estimated probability of liquefaction occurrence at various combinations of CSR7.5 and qc1N, allowing a probabilistic interpretation instead of relying solely on deterministic boundaries.

6.6 Validation

To validate the proposed probabilistic framework and highlight distinctions between competing models, multiple comparisons were carried out using transformed and untransformed explanatory variables. The performance of models based on non-normally distributed variables was evaluated to assess how variable transformations influence the fit and interpretability of the regression curves. The Bayesian model was further benchmarked against the traditional empirical method.

6.6.1 Comparisons using Non-Normally Distributed Explanatory Variables

To examine how the distribution of input variables affects logistic regression outcomes, the same dataset was used to develop probability models with two different combinations of predictors: (CSR7.5, qc1N) and (ln(CSR7.5), qc1N). Although neither pair strictly follows a normal distribution, the log-transformed CSR form has been previously adopted in liquefaction studies, such as by Lai et al. (1990), to address skewness in seismic loading data. Bayesian logistic regression results for different soil categories using both predictor sets are summarised in Tables 6.2 and 6.3, highlighting the influence of variable transformation on model coefficients. The results show that the Bayesian logistic regression models vary with the change of the explanatory variables and soil type. Different transformations of the same dataset can yield multiple valid regression functions, reducing the uniqueness of the model. As a result, some statistical limitations, such as transformation bias and model dependency, become unavoidable.

Table 6.2 Bayesian Logistic Regression Probability Models for Different Soils for (CSR7.5, qc1N) pair

Soil classification	Logit transformation of liquefaction probability	Number of cases
Sand	$\pi(X) = -0.196 + 1.397 * CSR7.5 - 1.151 * (qc1N)$	198
Sand Mixture	$\pi(X) = 1.121 + 2.050 * CSR7.5 - 0.270 * (qc1N)$	181
Silt Mixture	$\pi(X) = 2.071 + 2.645 * CSR7.5 - 0.072 * (qc1N)$	17

Table 6.3 Bayesian Logistic Regression Probability Models for Different Soils for (ln(CSR7.5), qc1N) pair

Soil classification	Logit transformation of liquefaction probability	Number of cases
Sand	$\pi(X) = -0.318 + 1.718 * \ln CSR7.5 - 1.401 * (qc1N)$	198
Sand Mixture	$\pi(X) = 0.959 + 1.657 * \ln CSR7.5 - 0.324 * (qc1N)$	181
Silt Mixture	$\pi(X) = 1.909 + 2.585 * \ln CSR7.5 + 0.128 * (qc1N)$	17

Each model employs the same dataset but varies in how the input variables are mathematically transformed, leading to observable differences in boundary shape and curvature. From Figure 6.9, we see that all three models converge or intersect near qc1N $\approx$ 25 and CSR7.5 $\approx$ 0.1, a region densely populated by data points. This intersection reflects a common boundary zone for classifying liquefaction and non-liquefaction conditions based on mid-range soil resistance and seismic demand. Beyond this convergence point, the curves diverge distinctly, particularly at higher qc1N values. Model 3, the original model produced by Bayesian logistic regression in Table 6.1, exhibits the sharpest curvature, rising steeply. Model 2 (green) tends to curve earlier and more dramatically, underpredicting resistance in the higher CSR regime, possibly due to the exponential scaling effect introduced by the logarithmic transformation of CSR. This model's boundary is tighter at low qc1N but becomes less representative at higher resistance levels. Model 1, which directly relates qc1N to CSR7.5 through a linear regression framework, indicated a positive correlation between the two parameters. While the model was able to broadly differentiate liquefied from non-liquefied cases, considerable overlap was observed at lower qc1N values, suggesting limitations in its ability to fully capture the separation between the two response categories. This highlights that, although Model 1 provides a straightforward representation of the qc1N–CSR7.5 relationships, its predictive capability is somewhat constrained in the lower resistance range.

From a geotechnical perspective, such variation in model curvature highlights the sensitivity of Bayesian logistic regression outcomes to variable transformation, which influences the interpretation of soil behaviour under cyclic loading. The transformations can capture nonlinearities, but may also introduce bias or unrealistic extrapolation if not grounded in physical behaviour. Larger curvature, as seen in Model 3, reflects better discrimination between liquefied and non-liquefied zones at extreme ends, offering a more conservative boundary, which is often preferable in seismic design to avoid underestimating hazard.

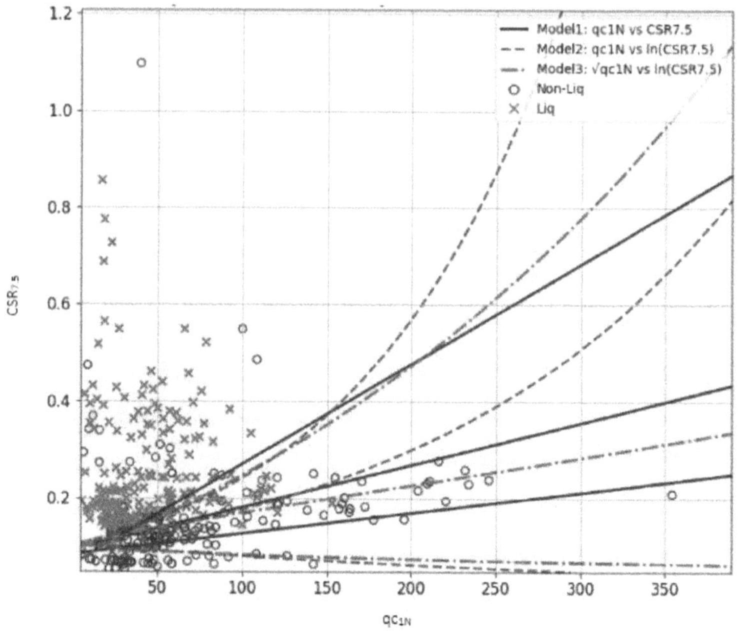

Fig. 6.9 The figure illustrates the P = 0.5 liquefaction boundaries derived from three logistic regression models using different formulations of explanatory variables:

6.6.2 Comparison of Developed Model with Other Empirical Models

To demonstrate the differences between the Bayesian probability model and earlier approaches, including an empirical model that incorporates the model proposed by Juang et al. (2002) in equation 6.15.

$$\pi(X) = 12.4259 + 3.9887\ln(\text{CSR } 7.5) - 0.0498\text{qc1N} \quad (6.15)$$

The two models were then evaluated using an independent dataset comprising 101 new case histories from Zhao et al. (2022), which includes records from the 2010 Darfield, 2011 Christchurch, 2011 Tohoku, 2012 Emilia, and 2012 Mirandola earthquakes.

Table 6.4. Performance of the two models

Model	Cutoff points	Accuracy	Precision	ROC AUC
Empirical model	0.4	0.72	0.67	0.85
Bayesian model	0.4	0.79	0.757	0.847
Empirical model	0.5	0.782	0.733	0.851
Bayesian model	0.5	0.80	0.796	0.847
Empirical model	0.7	0.782	0.786	0.851
Bayesian model	0.7	0.683	0.804	0.847

The performance of the Bayesian and empirical (fixed-coefficient) models was further evaluated using a quantitative comparison based on key classification metrics: accuracy, precision, and ROC AUC (The ROC AUC measures a model's ability to distinguish between two classes. It is the area under the ROC (Receiver Operating Characteristic) curve, where 0.5 indicates random performance and 1.0 indicates perfect discrimination, as summarised in Table 6.4. These metrics provide insight into the models' ability to correctly identify liquefied and non-liquefied cases under different cutoff thresholds, specifically, 0.4, 0.5, and 0.7.

a. Accuracy reflects the overall proportion of correctly classified cases (both liquefied and non-liquefied).
b. Precision indicates the proportion of predicted liquefied cases that were correctly important in reducing false positives.
c. ROC AUC (Receiver Operating Characteristic – Area Under Curve) summarises the model's overall discriminative ability, regardless of the threshold; values closer to 1 indicate better performance.

At a cutoff of 0.4, the Bayesian model outperforms the empirical model in both accuracy (0.79 vs. 0.72) and precision (0.76 vs. 0.67), with a comparable ROC AUC (~0.85). This suggests that the Bayesian model is more reliable in identifying liquefaction when a more sensitive threshold is used. At the standard 0.5 threshold, the Bayesian model outperforms the empirical model (0.80 vs 0.782), and the Bayesian model yields higher precision (0.80 vs. 0.73). This indicates that, even at the midpoint cutoff, the Bayesian model produces fewer false positives than the empirical model, offering more trustworthy classification of liquefaction-prone cases. However, at a higher cutoff of 0.7, the empirical model maintains high performance, while the Bayesian model shows a drop in accuracy (0.68), despite having slightly higher precision (0.80 vs. 0.79). This suggests that the Bayesian model becomes more conservative at higher thresholds, which may reduce false positives but at the cost of increased false negatives.

6.7 Conclusion

This study developed and validated a Bayesian logistic regression framework for liquefaction prediction using field-based CPT data, emphasising the effects of input variable transformations and model interpretability. The transformation of CSR7.5 and qc1n into ln(CSR7.5) and √qc1N, respectively, improved linearity and allowed clearer visualisation of probability contours in the liquefaction decision space. These transformations, grounded in geotechnical reasoning and statistical evaluation, revealed distinct interaction patterns between soil resistance and seismic demand across different soil types. Compared to traditional empirical methods, the Bayesian model offers notable advantages. It demonstrated higher accuracy and precision, particularly at lower and standard cutoff thresholds (e.g., 0.4 and 0.5), while maintaining comparable ROC AUC scores. Additionally, the probabilistic nature of the Bayesian approach allows for more flexible and interpretable decision boundaries that adapt to the underlying data structure and acknowledge uncertainty features not captured in fixed-coefficient models. The comparative analysis also highlights the sensitivity of logistic regression outcomes to variable transformations. The Bayesian model, while more responsive to curvature in the data space, avoids overfitting by incorporating prior distributions and full posterior inference. This leads to more robust generalisation across varying field conditions, as evidenced by its consistent performance on an independent validation set. Overall, Bayesian logistic regression emerges as a statistically principled and geotechnical interpretable tool for liquefaction hazard assessment. It not only improves predictive reliability but also enhances decision-making under uncertainty—an essential consideration in seismic geotechnical engineering.

References

Ang, Alfredo H-S., and Wilson H. Tang. "Probability concepts in engineering planning and design, vol. 2: Decision, risk, and reliability." *JOHN WILEY & SONS, INC., 605 THIRD AVE., NEW YORK, NY 10158, USA, 1984, 608* (1984).

Ayalew, Lulseged, and Hiromitsu Yamagishi. "The application of GIS-based logistic regression for landslide susceptibility mapping in the Kakuda-Yahiko Mountains, Central Japan." *Geomorphology* 65.1-2 (2005): 15-31.

Bhardwaj, Ravi Bhushan, and S. R. Chaurasia. "Use of ANN, C4.5 and Random Forest Algorithm in the Evaluation of Seismic Soil Liquefaction." *Journal of Soft Computing in Civil Engineering* 6.2 (2022): 92-106. https://doi.org/10.22115/SCCE.2022.314762.1380.

Carlin, B. P., and T. A. Louis. "Approaches to statistical inferences." *Bayesian methods for data analysis* 3 (2009).

Cheung, Raymond WM, and Wilson H. Tang. "Realistic assessment of slope reliability for effective landslide hazard management." *Geotechnique* 55.1 (2005): 85-94. https://doi.org/10.1680/geot.2005.55.1.85

Dai, F. C., and C. F. Lee. "A spatiotemporal probabilistic modelling of storm-induced shallow landsliding using aerial photographs and logistic regression." *Earth Surface Processes and Landforms: The Journal of the British Geomorphological Research Group* 28.5 (2003): 527-545.

Garson, D. (2003). "Logistic regression." http://www2.chass.ncsu.edu/garson/pa765/logistic.htm.

Gelfand, Alan E., and Adrian FM Smith. "Sampling-based approaches to calculating marginal densities." *Journal of the American statistical association* 85.410 (1990): 398-409.

Gelman, Andrew, and Cosma Rohilla Shalizi. "Philosophy and the practice of Bayesian statistics." *British Journal of Mathematical and Statistical Psychology* 66.1 (2013): 8-38. **https://doi.org/10.1111/j.2044-8317.2011.02037.x**

Givens, Geof H., and J. A. Hoeting. "Computational Statistics, John Wiley&Sons." *Inc., Hoboken, New Jersey* (2013).

Goh, Anthony TC, and S. H. Goh. "Support vector machines: their use in geotechnical engineering as illustrated using seismic liquefaction data." *Computers and Geotechnics* 34.5 (2007): 410-421. https://doi.org/10.1016/j.compgeo.2007.06.001

Goh, Anthony TC. "Neural-network modeling of CPT seismic liquefaction data." *Journal of Geotechnical engineering* 122.1 (1996): 70-73. **https://doi.org/10.1061/(ASCE)0733-9410(1996)122:1(70)**

Goh, Anthony TC. "Seismic liquefaction potential assessed by neural networks." *Journal of Geotechnical engineering* 120.9 (1994): 1467-1480. **https://doi.org/10.1061/(ASCE)0733-9410(1994)120:9(1467)**

Hastings, W. Keith. "Monte Carlo sampling methods using Markov chains and their applications." (1970): 97-109.

Jha, Sanjay K., and Kiichi Suzuki. "Reliability analysis of soil liquefaction based on standard penetration test." *Computers and Geotechnics* 36.4 (2009): 589-596. https://doi.org/10.1016/j.compgeo.2008.10.004

Jiang, S. H., I. Papaioannou, and D. Straub. "Bayesian updating of slope reliability in undrained clay with vane shear test data." *14th International Probabilistic Workshop*. Springer, Cham, 2017.

Johnson, N. L., and T. W. Anderson. "Introduction to Multivariate Statistical Analysis." Journal of the Royal Statistical Society. Series A (General), vol. 121, no. 4, 1958, p. 482. Crossref, https://doi.org/10.2307/2343317.

Juang, C. H., et al. "CPT-Based Liquefaction Analysis, Part 1: Determination of Limit State Function." Géotechnique, vol. 50, no. 5, Oct. 2000, pp. 583–92. Crossref, https://doi.org/10.1680/geot.2000.50.5.583.

Juang, C. Hsein, et al. "Assessing Probability-Based Methods for Liquefaction Potential Evaluation." Journal of Geotechnical and Geoenvironmental Engineering, vol. 128, no. 7, July 2002, pp. 580–89. Crossref, https://doi.org/10.1061/(asce)1090-0241(2002)128:7(580).

Kohestani, V. R., M. Hassanlourad, and A. J. N. H. Ardakani. "Evaluation of liquefaction potential based on CPT data using random forest." *Natural Hazards* 79.2 (2015): 1079-1089. https://doi.org/10.1007/s11069-015-1893-5

Lai, Sheng-Yao, et al. "Discriminant Model for Evaluating Soil Liquefaction Potential Using Cone Penetration Test Data." Journal of Geotechnical and Geoenvironmental Engineering, vol. 130, no. 12, Dec. 2004, pp. 1271–82. Crossref, https://doi.org/10.1061/(asce)1090-0241(2004)130:12(1271).

Lai, Sheng-Yao, et al. "Discriminant Model for Evaluating Soil Liquefaction Potential Using Cone Penetration Test Data." Journal of Geotechnical and Geoenvironmental Engineering, vol. 130, no. 12, Dec. 2004, pp. 1271–82. Crossref, https://doi.org/10.1061/(asce)1090-0241(2004)130:12(1271).

Lee, S. A. R. O. "Application of logistic regression model and its validation for landslide susceptibility mapping using GIS and remote sensing data." *International Journal of remote sensing* 26.7 (2005): 1477-1491.

Lee, Saro, and Biswajeet Pradhan. "Landslide hazard mapping at Selangor, Malaysia using frequency ratio and logistic regression models." *Landslides* 4.1 (2007): 33-41.

Lee, Saro. "Application of likelihood ratio and logistic regression models to landslide susceptibility mapping using GIS." *Environmental Management* 34 (2004): 223-232.

Liang, Wan-jie, et al. "Assessment of debris flow hazards using a Bayesian Network." *Geomorphology* 171 (2012): 94-100.

Liao, Samson S. C., and Robert V. Whitman. "Overburden Correction Factors for SPT in Sand." Journal of Geotechnical Engineering, vol. 112, no. 3, Mar. 1986, pp. 373–77. Crossref, https://doi.org/10.1061/(asce)0733-9410(1986)112:3(373).

Liao, Samson S. C., et al. “Regression Models For Evaluating Liquefaction Probability.” Journal of Geotechnical Engineering, vol. 114, no. 4, Apr. 1988, pp. 389–411. Crossref, https://doi.org/10.1061/(asce)0733-9410(1988)114:4(389).

Liao, Samson S. C., et al. “Regression Models For Evaluating Liquefaction Probability.” Journal of Geotechnical Engineering, vol. 114, no. 4, Apr. 1988, pp. 389–411. Crossref, https://doi.org/10.1061/(asce)0733-9410(1988)114:4(389).

Liao, Samson SC, Daniele Veneziano, and Robert V. Whitman. "Regression models for evaluating liquefaction probability." *Journal of Geotechnical Engineering* 114.4 (1988): 389-411. **https://doi.org/10.1061/(ASCE)0733-9410(1988)114:4(389)**

Menard, Scott. Applied Logistic Regression Analysis. SAGE Publications, Inc., 2002. Crossref, https://doi.org/10.4135/9781412983433.

Metropolis, Nicholas, et al. "Equation of state calculations by fast computing machines." *The journal of chemical physics* 21.6 (1953): 1087-1092.

ohnson, R. A., and D. W. Wichern. “Applied Multivariate Statistical Analysis.” Biometrics, vol. 54, no. 3, Sept. 1998, p. 1203. Crossref, https://doi.org/10.2307/2533879.

Oommen, Thomas, Laurie G. Baise, and Richard Vogel. "Validation and application of empirical liquefaction models." *Journal of geotechnical and geoenvironmental engineering* 136.12 (2010): 1618-1633. **https://doi.org/10.1061/(ASCE)GT.1943-5606.0000395**

Pal, Mahesh. "Support vector machines-based modelling of seismic liquefaction potential." *International Journal for Numerical and Analytical Methods in Geomechanics* 30.10 (2006): 983-996. **https://doi.org/10.1002/nag.509**

Robertson, P. K. “Soil Classification Using the Cone Penetration Test.” Canadian Geotechnical Journal, vol. 27, no. 1, Feb. 1990, pp. 151–58. Crossref, https://doi.org/10.1139/t90-014.

Seed, H. Bolton, and Izzat M. Idriss. “Simplified Procedure for Evaluating Soil Liquefaction Potential.” Journal of the Soil Mechanics and Foundations Division, vol. 97, no. 9, Sept. 1971, pp. 1249–73. Crossref, https://doi.org/10.1061/jsfeaq.0001662.

Sherri, Marwan, et al. "A Comparison of Multiple Markov Chains Algorithms for Bayesian Updating." *2021 International Conference on Electrical, Computer and Energy Technologies (ICECET)*. IEEE, 2021. 10.1109/ICECET52533.2021.9698446

Straub, Daniel, and Iason Papaioannou. "Bayesian updating with structural reliability methods." *Journal of Engineering Mechanics* 141.3 (2015). doi:10.1061/(ASCE)EM.1943-7889.0000839

Toprak, Selcuk, et al. "CPT-and SPT-based probabilistic assessment of liquefaction." *Proc., 7th US–Japan Workshop on Earthquake Resistant Design of Lifeline Facilities and Countermeasures against Liquefaction*. Buffalo, NY: Multidisciplinary Center for Earthquake Engineering Research, 1999.

Youd, T. Leslie, and Izzat M. Idriss. "Liquefaction resistance of soils: summary report from the 1996 NCEER and 1998 NCEER/NSF workshops on evaluation of liquefaction resistance of soils." *Journal of geotechnical and geoenvironmental engineering* 127.4 (2001): 297-313.

Zhang, et al. "An improved method for calculating slope length (λ) and the LS parameters of the Revised Universal Soil Loss Equation for large watersheds." *Geoderma* 308 (2017): 36-45. https://doi.org/10.1016/j.geoderma.2017.08.006.

Zhao, Zening, et al. “CPT-Based Fully Probabilistic Seismic Liquefaction Potential Assessment to Reduce Uncertainty: Integrating XGBoost Algorithm with Bayesian Theorem.” Computers and Geotechnics, vol. 149, Sept. 2022, p. 104868. Crossref, https://doi.org/10.1016/j.compgeo.2022.104868

CHAPTER 7

Summary and Conclusions

7.1 Summary

This thesis presents four distinct studies applying Bayesian and machine learning techniques to key challenges in geotechnical engineering:

a) Bayesian Multilevel Modelling (MLM) for Shear Wave Velocity (Vs) Prediction: Investigates regional variability in the correlation between Vs and SPT blow count (N) for sandy soils using a Bayesian MLM approach. A stable slope across regions with varying intercepts is confirmed.
b) Bearing Pressure Prediction using AdaBoost-BOA and GBoost-BOA Models: Develops ensemble learning models, optimised through Bayesian Optimisation, to predict bearing pressure of shallow foundations on clayey soils using PLT data. GBoost-BOA significantly outperforms others in prediction accuracy and generalisability.
c) Hybrid Bayesian Network for Soil Groutability Prediction: Proposes a hybrid BN model combining discrete and continuous variables for classifying groutable soils. The model minimises information loss, outperforms traditional methods, and prioritises high recall for successful grouting identification.
d) Bayesian Logistic Regression for Liquefaction Prediction using CPT Data: Constructs a probabilistic model to estimate liquefaction potential using transformed CPT-derived variables. The Bayesian framework offers enhanced interpretability and uncertainty quantification over conventional logistic models.

7.2. Conclusions

a) A Bayesian multilevel modelling (MLM) framework was effectively employed to calibrate the relationship between shear wave velocity (Vs) and standard penetration test (SPT) blow counts (N) in sandy soils. The model confirmed a stable slope across different geographic regions, while capturing intercept variations attributed to local soil conditions and measurement uncertainties.

b) The transformation to natural logarithmic space (ln(Vs) vs. ln(N)) successfully linearised the Vs–N relationship, allowing the MLM to express both within-region and between-region variability through hierarchical structures. This transformation preserved physical interpretability and facilitated cross-site comparison.
c) The Bayesian paradigm enhanced model interpretability by enabling posterior distribution estimation for all parameters. It also provided credible intervals for both slope and intercept, allowing practitioners to make decisions under uncertainty, a feature lacking in traditional regression approaches.
d) Partial pooling in the MLM prevented overfitting to sparse regional datasets while still allowing regional characteristics to influence the model, leading to improved predictive accuracy and generalizability.
e) The approach quantified three layers of uncertainty: within-site variability, between-site variability, and uncertainty in global parameters.
f) The Resulting model demonstrated strong performance on unseen regions or new soil clusters.
g) The calibrated model and framework can be extended to predict shear wave velocity (Vs) from standard penetration test (N) values for new, unseen data in sandy soils. Additionally, the framework can be adapted for Vs–N correlations in other soil types or for developing other geotechnical correlations, making it a reusable and scalable modelling tool.
h) Predicting bearing pressure of spread footings on clayey soils, ensemble learning models (AdaBoost and GBoost) optimised via Bayesian Optimisation Algorithm (BOA) demonstrated superior performance compared to standalone models and artificial neural networks (ANNs). These models efficiently handled complex, nonlinear interactions between multiple soil and footing parameters.
i) The use of a large and diverse plate load test dataset reduced overfitting and increased the applicability of the model to various site conditions.
j) The hybrid GBoost-BOA model, in particular, achieved exceptional predictive metrics ($R^2 > 0.98$, RMSE ~20 kPa, MAE ~13 kPa), with consistent performance across both training and testing datasets. This consistency, combined with external validation and SHAP-based feature importance analysis, confirmed the model's robustness and interpretability.
k) The use of Bayesian optimisation was crucial for efficient and reliable hyperparameter tuning. BOA provided computationally efficient exploration of the hyperparameter space, significantly reducing training times while maximising generalisation performance.
l) A Hybrid Bayesian Network (BN) model was developed to predict soil groutability, overcoming the limitations of traditional discrete BNs by directly incorporating both continuous and discrete variables.
m) The Hybrid BN model utilised Markov Chain Monte Carlo (MCMC) methods for parameter learning and structure discovery, maintaining predictive accuracy even in the presence of class imbalance and limited datasets. This adaptability is particularly

valuable in real-world geotechnical applications, where complete and balanced datasets are rare.

n) Discretisation often leads to information loss in geotechnical data; avoiding it preserves natural soil variability and transition behaviour.
o) The model required significantly less memory and computation time compared to discrete BN, thanks to more efficient parameter storage.
p) Comparative analysis against Discrete BN, Naïve Bayes, and Support Vector Machines (SVM) showed that the Hybrid BN consistently outperformed other models in terms of recall, accuracy, and interpretability. High recall ensured that groutable samples were not misclassified, a critical requirement for safe grouting design.
q) The prediction of groutability is heavily dependent on maximising recall, as it is crucial to identify as many groutable soil samples as possible under varying soil and grout conditions. This emphasis on recall ensures that potentially groutable samples are not overlooked, even if it results in a higher rate of false positives. Consequently, the BN model tends to favour recall-focused predictions, which may skew its outcomes toward detecting groutable samples more effectively. This trade-off aligns with the primary objective of groutability analysis, where minimising the risk of missed groutable cases outweighs concerns regarding false positive identifications.
r) The model also demonstrated scalability and adaptability, as it can be updated incrementally with new data without reconfiguring the network structure, a key advantage for dynamic geotechnical systems and decision-support applications.
s) A Bayesian logistic regression model was developed for liquefaction potential prediction based on Cone Penetration Test (CPT) data. Statistically informed transformations of the input parameters ($\sqrt{q_{c1}N}$ and ln(CSR7.5)) improved the linearity of the predictor space and enhanced the interpretability of the resulting logistic surface.
t) The Bayesian model provided probability distributions over predicted outcomes, allowing for a transparent estimation of uncertainty in liquefaction classification. This is particularly important in seismic engineering, where false negatives can lead to catastrophic failure.
u) Compared to traditional logistic regression, the Bayesian approach achieved higher precision at standard decision thresholds (e.g., $p = 0.5$), more adaptable decision boundaries, and better generalisation on independent datasets.
v) All models were developed using published literature data and validated on entirely separate, unseen datasets also sourced from the literature. The exclusive use of published data enhances transparency and reproducibility, allowing future researchers to verify and extend the work.
w) Across all chapters, a consistent theme is the shift from black-box modelling to transparent, explainable models. By focusing on parameter distributions, variable importance, and prior knowledge integration, the thesis bridges the gap between statistical power and engineering practice.
x) The thesis also demonstrates that with thoughtful model design, existing field data can be leveraged to reduce the need for costly new testing. Models that learn from regional

datasets or historical trends can support faster, more cost-effective engineering decisions, particularly during early design phases or in data-scarce environments.

y) Overall, the thesis promotes a paradigm shift in geotechnical engineering towards probabilistic, interpretable, and data-conscious modelling practices. It encourages engineers to treat uncertainty not as a nuisance, but as an opportunity for better-informed, more resilient design.

7.3 Probable Areas of Application

- Regional seismic studies
- Development of Vs profiles for ground response analysis
- Rapid estimation of Vs in preliminary site investigations
- Foundation design in soft or clayey soils
- Predictive modelling for settlement and bearing capacity in infrastructure projects
- Alternative to costly or unavailable full-scale PLTs
- Risk-based decision-making for permeation grouting in tunnelling and underground construction
- Pre-injection grout design for diverse geologic conditions
- Probabilistic liquefaction hazard mapping

7.4. Limitations of the Present Study

- Assumes a fixed slope across regions, which might not hold for drastically different geological settings.
- Limited to sandy soils, cannot be extrapolated to silts or clays without further calibration.
- Dataset, though diverse, may still not encompass all the variability present in global geotechnical conditions.
- Model performance may degrade when extrapolated to soil types or loading conditions not represented in the training data.
- Gaussian assumption for continuous variables may misrepresent certain geotechnical distributions in the BN model.
- Lack of interpretability in some BN substructures due to complex interactions may limit practical use for non-specialists.
- Sensitivity to input variable transformation necessitates expert judgment.
- Model performance is constrained by the quality and completeness of the CPT database.

7.5 Future Scope

- Extend MLM to include mixed soils and layered profiles.
- Integrate additional geotechnical parameters (e.g., fines content, relative density) to improve regional predictions.
- Develop global ML frameworks using larger, open-access PLT datasets.
- Explore integration with physics-informed ML to retain theoretical grounding.
- Incorporate copula functions or non-Gaussian distributions for more realistic modelling of soil-grout interactions.
- Test hybrid BN models on field-scale grouting projects for deployment-ready calibration.
- Introduce spatial Bayesian models to account for spatial correlation in liquefaction occurrence.
- Exploration of hybrid physics-informed data models, which might improve soil performance.

www.ingramcontent.com/pod-product-compliance
Lightning Source LLC
LaVergne TN
LVHW051923051126
851857LV00021B/2108
* 9 7 9 8 2 4 0 8 0 6 7 5 9 *